COLLECTING NATURE

A History of the Herbarium and Natural Specimens

COLLECTING NATURE

A History of the Herbarium and Natural Specimens

CLIVE ASLET

SVANTE HELMBAEK TIRÉN

Translation from Swedish: Fern Scott Olsson

BOKFÖRLAGET STOLPE

AXEL AND MARGARET AX:SON JOHNSON FOUNDATION FOR PUBLIC BENEFIT

Contents

Preface

The World Heritage Site of Engelsberg Ironworks in the county of Västmanland, Sweden, is home to a large variety of historical collections. While some of them are directly linked to centuries of iron making, others reflect the cultural, scientific and artistic interests of the various families that owned and operated the Engelsberg works. These collections contain a rich spectrum of knowledge and subjects that continue to illuminate the historical significance of the site and its industrial past.

This book, which explores the natural history collection created by the Timm family in the nineteenth century, is part of a series that places these collections in a wider context, viewed from different perspectives by scholars and experts. Books on the history of subjects as diverse as wallpapers and cocklestoves have already been published, and other books will follow to shed more light on Engelsberg and its manifold roles as a heritage site.

Through a collaboration with Bokförlaget Stolpe, this material is presented in a format whereby text and illustrations aim to inspire interest in and awareness of these unique collections, for both a Swedish and an international audience.

Axel and Margaret Ax:son Johnson Foundation for Public Benefit is a private foundation with the primary purpose of promoting scientific and scholarly research in general. Founded in 1947 by the owner of the Nordstjernan Group, Consul-General Axel Ax:son Johnson (1876–1958), and his wife Margaret (1887–1966), its focus today is on humanities and social sciences.

Kurt Almqvist
President

Axel and Margaret Ax:son Johnson Foundation for Public Benefit

The Timm herbarium at Engelsberg in context

CLIVE ASLET

O ne of the many wonders of Engelsberg, the Swedish World Heritage Site, is the herbarium that the Timm family assembled there in the nineteenth century and left in the country house. Although large, it is by no means enormous when compared to the herbaria that exist in national museums around the world; nor are the thousands of exhibits particularly remarkable botanically. And yet, it is an extraordinary entity. Everyone knows from childhood that pressed flowers are fragile, but – partly through careful stewardship, partly by chance – those in the Timm Herbarium have survived for two centuries exceptionally intact, the colours of their blooms still, in some cases, fresh.

Arranged with an eye to symmetry and other aesthetic considerations, the long-dead plants retain something of their living beauty, although it is not as we usually find it in nature; stalks have been reduced to skeletons, leaves to a web of veins, giving them the eerie and angular fascination of Javanese shadow puppets. Open the dark-stained pine cabinets that contain these samples and the aroma of long-vanished summers – the straw-bale smell of dried vegetation – envelops you. It is as though the sunshine of past centuries has been bottled and, like wine, transmuted into a different medium. Birds' eggs, minerals, a few stuffed birds and some cases of beetles, their wings still iridescent, exist alongside the plants. Notes are attached to the samples, written in the hand of the person who found or acquired them.

Today, herbaria are not widely kept and the subject can seem somewhat niche. Thousands of those collected by families like the Timms across Europe must have been thrown out in the course of the twentieth century, as part of the great bonfire of Victorian clutter. This makes the Timm Herbarium a rare survivor. Before photography, however, there was no alternative to pressing and drying plant material for those who wanted to study it seriously; even the paintings of botanical artists, which could be lifelike as well as ravishing, do not contain so much information as the thing itself when carefully preserved. To this day, the study of the actual specimen is critical, not least to extract DNA and other data that have given new relevance to old collections.

Moreover, cameras were not easy to carry on expeditions, in the age of film. Early equipment was cumbersome, film was expensive and it could be ruined by exposure to light before prints were made. Some gentlemen and many ladies, like the Georgian botanical illustrator Elizabeth Blackwell, could draw to

Although they are nearly 200 years old, many of the herbarium's plants are extremely well preserved. Even today, one is struck by their beauty, and how special it is to botanise among the collection's thousands of sheets of flowers, herbs, grasses and other curiosities.

Plate 245.

Male-Piony.

1. *Flower.*
2. *Seed Veſsel.*
3. *Seed Veſsel open.*
4. *Seed.*

Paeonia mas.

Eliz. Blackwell delin. sculp. et Pinx.

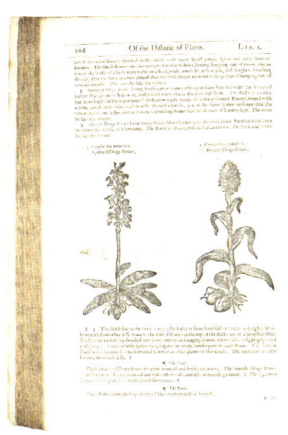

Early botanical writings and publications had very limited scope to reproduce the plants, especially when it came to illustrations in colour. Here we see a page from *The Herball or Generall Historie of Plants* by John Gerard, first published in 1597.

Opposite page: Elizabeth Blackwell (1707–1758) was a Scottish botanical illustrator and author, best known for her work, *A Curious Herbal*, which was published in the 1730s. The book contains over 500 depictions of different plants of diverse utility. This illustration is of a brilliant red peony, with additional details.

a high standard, but they worked slowly. People without such artistic gifts, or with little time on their hands, needed herbaria. Certainly this was true of the plant hunters who explored wild areas in sometimes inhospitable conditions.

Rarity is not the only thing that makes the Timm Herbarium important. We live in a changing world. Species adapt to new conditions, or become extinct, driven by a multitude of factors, including – as we know so well – the climate. To science, any plant that has been picked and labelled with the time and place that it was found has value. It forms a benchmark by which comparisons can be made. Herbaria may be an exercise in desiccation because their contents have been deliberately dried out, but they come alive again under the lens of the photographer and the eye of science.

The Timm Herbarium belongs to a grand tradition, the origins of which have been traced to the Renaissance. It is tempting to believe they could have been more distant than that. The process of preserving plants by pressing them between two leaves of paper is simple; children do it to this day, by putting daisies and buttercups inside books. (Once the moisture has been squeezed out of the samples, they become stable and, if properly cared for, do not change, except in colour.) In the Middle Ages, some monks spent their days surrounded by herbs of all kinds, often dried, to use in medicine. Others accurately depicted living plants in the herbals, or plant encyclopedias, that they made; one imagines that some of the models they copied would have been pressed to preserve them. Paper was not produced commercially in England until the late sixteenth century, but, even so, might not some monasteries have kept a chest of dried flowers for instruction? However, if monks did keep herbaria, they have disappeared and we have no reference to them. Barbara M Thiers, in her book *Herbarium*, states that the first herbarium was created by Luca Ghini, a physician born in Bologna around 1490.[1] Ghini taught medicine at the University of Bologna, lecturing on 'simples' or medicinal herbs.

Readers of Shakespeare will know that elaborate folklore was associated with plants in the sixteenth and seventeenth centuries. Fairies were supposed to cradle their babies in the flowers of thyme, while hellebore was believed to have the power to repel witches. Plants also provided some of the few efficacious treatments for sickness available at the time. Consequently, the study of plants – what we would now call botany – was highly respected and everyone had a practical interest in sharing the knowledge. Great noblemen such as Lord Burghley were proud to finance the printing of herbals – illustrated books about plants and their properties. However, the woodcuts in works such as John Gerard's illustrated *Herball, or Generall Historie of Plantes*, first published in 1597 and the basis for much of England's understanding of plants in the course of the next century, were crude. A herbarium which contained examples of actual plants in dried form provided a better overview and far more information that complemented the more widely held knowledge of plants.

Another tradition that emerged in the sixteenth century was the cabinet of curiosities or *Wunderkammer*. Great personages and humanist scholars displayed their culture, learning and wealth by assembling marvels of natural history, geology, exquisite craftsmanship and the exotic – sometimes almost a sort of equivalent to the collections of saints' relics that were amassed by the Church. Narwhals' tusks, baroque pearls, coconut shells, nautilus shells, painted miniatures and other wonders (some, like the relics, manufactured for the purpose of sale) were displayed in small rooms (cabinets) and shown to the owners' closest friends. These objects from the natural world could be given elaborate metalwork settings in the manner of jewels. They both delighted the senses and stirred the curiosity of the beholders, whose interest might become scientific.

The artist Giuseppe Arcimboldo (approx. 1527–93) painted portraits that pieced together animals and parts of plants, among other things, to portray people. Here the Holy Roman Emperor Rudolf ll is depicted as Vertumnus, the god of the gardens and seasons in Roman mythology.

Opposite page: The artist Joris Hoefnagel (1542–1600) produced some of the most playful paintings of plants and wildlife of his time. Here is his depiction of a large flower in full bloom, with a number of insects included to create an image that is both festive and fascinating.

Following pages: The Danish physician Ole Worm (1588–1654) created this cabinet of curiosities. The room's natural objects included plants, minerals and various animal species, alongside antiques and works of art.

Every prince, potentate, prelate and aristocrat had a cabinet, the most magnificent being that of the Holy Roman Emperor Rudolf II of Bohemia. To say that Rudolf collected everything is not wholly hyperbolic. He intended that his collection should be a microcosm of the world as a whole, containing all that was rare, beautiful, wonderful and exquisite – 'the three kingdoms of nature and the works of man' that Wendell E Wilson described in the catalogue to an exhibition – 'Masterpieces of the Mineral World' – at the Houston Museum of Natural Science in 2004. As well as commissioning works of art from Europe's greatest masters, Rudolf surrounded himself with examples of natural history, often bizarre – or, in the language of the time, baroque. A lion and a tiger prowled Prague Castle, part of the zoo and botanical garden that he created. Works of dazzling craftsmanship included paintings on semi-precious stones, landscapes made out of *pietre dure* or coloured stones and objects thought to have magical properties, such as rhinoceros horn and stones from the stomachs of animals; Rudolf had a taste for the occult as well as an insatiable love of jewels, as can be seen today from those in the Kunsthistorisches Museum in Vienna. Rudolf also patronised scientists, bought scientific instruments, had himself painted by Giuseppe Arcimboldo as Vertumnus, the Roman god of seasons, change and plant growth – his features are represented by vegetables, fruit and flowers, as a reflection on the universal principles that govern nature – and owned a natural history collection. The last included 100 species of colourful butterflies and a herbarium.

The butterflies were lost after Rudolf's death, perhaps during the sack of Prague by Swedish forces during the Thirty Years' War. We have an idea of the herbarium, however, from the *trompe-l'oeil* paintings of Joris Hoefnagel; illusionism was itself thought of as a branch of science. (Hoefnagel's services were, of course, expensive. Poorer collectors who sought to record plants in two-dimensional form could do so by inking them and taking impressions on paper; the results, however, must have been crude compared to the intricate detail we see in the dried plants of the Timm Herbarium and others like it.)

AMORIS MONVMENTŨ MATRI CHARISS:
GEORGIVS HOEFNAGLIVS. D. Aº 89.

MU
WORM
HIS
LUGD · B.
EX OFFICINA
Acad Ty

LICA

METALLA MINERALIA

P.

LAPIDES

LAPIDES

LAPIDES

RBINATA CONCHILIA MARIANA

SUCCI

FRUCTUS

SULPHURA

LIUM PARTES CONCHILIATA

SEMINA

LIGNA

SALIA

VARIA

CORTICES

TER

HERBÆ

I
ANI
RIA
VORUM
VIRIANA
1655.

RADICES

Wingendorff fc.

The riches of Rudolf's collection were a statement both of his intellectual reach and his power as an emperor. His Hapsburg successors did not maintain his *Wunderkammer*, which must have come to seem antiquated, if not ridiculous, but they shared his passion for the natural world. Gardening, flowers and collecting butterflies, along with the keeping of herbaria, were practically a family inheritance, until the collapse of the Austro-Hungarian monarchy in 1918. It may seem a long way from the courts of Prague and Vienna to rural Sweden, but a distant echo of the princely cabinet of curiosities can be heard at Engelsberg, where the diversity of the Timm collections – embracing birds' eggs, butterflies and minerals, as well as samples of dried plants – can be seen as deriving from that older tradition. To the modern eye, theirs was an *omnium gatherum* approach, whereby they collected anything

Sir Hans Sloane (1660–1753), portrayed by Stephen Slaughter with a botanical illustration in his hands.

Opposite page: The artists Jan Brueghel the Elder and Hieronymus Francken II cooperated in the creation of this painting of the Archduke Albert VII of Austria (1559–1621) with his wife Isabella (1566–1633) visiting a magnificent collection. The walls are adorned with spectacular paintings, natural history objects are lying on the table, and living plants and animals are included to complete the scene.

within the field that struck them as interesting, because a wide range of science was accessible to a well-read amateur.

One of the greatest of herbaria from the years around 1700 was formed by the Irish doctor Sir Hans Sloane. Sloane was born in 1660, the year in which Charles II was restored to the throne; it was also the year that the Royal Society – formally the Royal Society of London for Improving Natural Knowledge – was founded, to promote the new or experimental philosophy being pursued by 'plain, diligent observers' of the world around them (to quote Thomas Sprat, the Royal Society's first historian[2]). It was an elite club, where scholars, who generally regarded themselves as gentlemen, associated with other notable figures, such as Samuel Pepys, and aristocrats who took an interest in science. Subjects of enquiry ranged from chemistry and microscopes to bread-making, extraordinary occurrences – such as the shoal of fish that apparently fell from the heavens during a rainstorm in Kent – and inventions proposed by the fellows, such as a cart worked by artificial legs, rather than wheels. Anything in the natural world was of interest in that age of curiosity, including plants and their properties. Botanical questions were avidly discussed at Royal Society meetings, and, after Sloane's death in 1753, 50 plants were sent from the Chelsea Physic Garden each year, under the terms of his will.

Sloane's first fortune was made in Jamaica, where he served as personal physician to the island's governor, the Duke of Albemarle. While there, he discovered what he could about the natural history of the West Indies, particularly its 'extraordinary' features. On his return to England, he published his observations in the *Philosophical Transactions of the Royal Society* and later as a book, *A Voyage to [...] Jamaica, with the natural history [of the island]*. He married a wealthy widow, whose husband had been a plantation owner (making Sloane a figure of opprobrium to those who wish to decolonise history). His medical career also prospered; he became physician-extraordinary to Queen Anne and bought the manor of Chelsea, where his name would be remembered in Sloane Square, Sloane Street, Sloane Avenue and Hans Place. It also enabled him to pursue a gargantuan and omnivorous appetite for collecting. The beautiful and the curious, the useful and the bizarre, the natural and the artificial, the incredibly old, the new but exquisite, as well as (in Horace Walpole's opinion) 'hippopotamuses, sharks with one ear, and spiders as big as geese' – all found their way into Sloane's houses in Bloomsbury Place and Chelsea, along with fossils, zoological specimens, anatomical oddities, antiquities, prints, drawings, coins, books and manuscripts. When Sloane died in 1753, Parliament voted to grant £20,000 to buy the collection, and it became the foundation of the British Museum.

As described so far, Sloane's collection might be conceived as a gigantic *Wunderkammer*, comprising 80,000 objects. But at the same time – and following, perhaps, the same instinct for assembling knowledge – Sloane also kept a herbarium. This survives in the Natural History Museum in London and is the subject of *The Collectors: creating Hans Sloane's extraordinary herbarium*, edited by Mark Carine – an essential resource, since the herbarium samples are too fragile to handle.[3] Sloane's herbarium comprised 120,000 specimens. This is small in comparison to the size of the Natural History Museum's herbarium today, which runs to five million specimens, but without parallel in the period it was made – from the 1680s until 1753, when Sloane died.

While the Timm Herbarium is on nothing like the scale of the Sloane Herbarium, there is a parallel with this and other herbaria in that both were the work of many hands. In the nineteenth century, when the Timms corresponded with the botanist Elias Fries and must at least have known of the work of their fellow iron masters, the Grill family at Söderfors, botanists – whether scholars or dedicated amateurs like the Timms – swapped plants with one another, as part of the exchange of scientific ideas that had characterised the Enlightenment. So it was for Sloane. A charming example is presented by Mrs Mary Lisle of Crux Eaton in Hampshire, who contributed only half a dozen plants, but corresponded with Sloane on how 'to preserve the Couler of Plants' – and the violet of her larkspur remains exceptionally bright. Altogether some 300 people contributed to the great work.

Like Sloane's own career, the contents of his herbarium reflect Britain's nascent empire and growing trade around the world, since foreign specimens were acquired by people who had a reason to travel, such as ships' surgeons, missionaries and employees of the East India Company. Sloane's friend William Courten supplied such travellers with scissors, pins and the other materials necessary to gather specimens. Despite being plagued by debts inherited from his father and grandfather, Courten assembled so great a collection that he displayed it to the public in a museum that occupied ten rooms in the Middle Temple. Ladies of the court and other visitors could gather there to admire shells, insects, medals, minerals, precious stones, objects in amber and dried fish and plants, some of them contributed by Sloane. Courten left Sloane the collection on his death, estimating it to be worth the large sum of £7,000 or £8,000.

An even bigger collection was formed by Sloane's friend James Petiver, the London apothecary. Like Courten, Petiver enlisted the help of travellers in foreign lands to 'make Collections for me of whatever Plants, Shells, Insects, &c. they shall meet with', as he wrote in the 1690s. After Petiver's death in 1718, Sloane, who had been a pall bearer at his funeral, praised him in the introduction to the second volume of his *A Voyage to Jamaica* as a man of profound 'Understanding in Natural History', who had assembled a 'greater Quantity' of specimens 'than any Man before him'. Sloane bought his herbarium; even at the end of Sloane's life, when his herbarium had grown to epic proportions, Petiver's specimens accounted for a third of the contents. For all his admiration of Petiver, Sloane criticised the lack of care

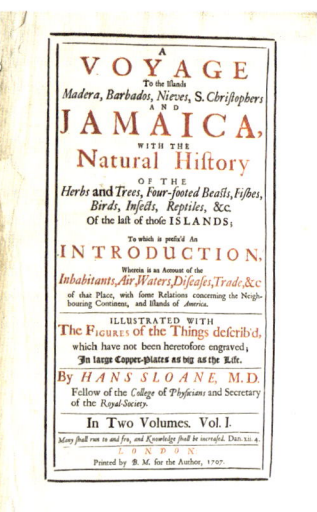

The title page of *A Voyage to Jamaica* from 1725, one of the publications that established Hans Sloane as an authority in the field of natural history. The book gave Europeans fresh insights into a new world, where every new species and discovery became important parts of the puzzle in an expanding area of knowledge.

Opposite page: In this letter from 1737 written in Latin, Hans Sloane thanks Carl Linnaeus for the gift of books he had received. One of them was *Flora Lapponica*, an account of the plants of Lapland, which enabled Sloane to note that several of the species described in the book could also be found in England and Ireland.

Vir eruditissime,

Quòd urbanissimis tuis literis, Martii 2° 1736, et Jan: 21: 1737
datis, responsa non maturius expediverim, pudore non levi
perfundor. Sed si negotiorum farragini, valetudinique haud
stabili; non verò socordiæ aut ingrato animo, moram hanc
ascribere benignè velis, veram attinges silentii causam. Jam
aliquantulum otii nactus, sinceras Tibi grates ago, tum ob
epistolas tuas acceptissimas, tum ob operas, quibus historiam
naturalem ampliare, bibliothecamque meam ditare dignatus
es. Hac quidem cum voluptate perlegi: Flora verò Lapponica
specialim mihi tantoperè arridet, ut maximè cupiam cæteras
illius regionis partes Historiæ naturalis intueri tuâ exaratas
manu, publicæque luci datas.

Tuam Æstri Lapponum accuratam valdè descriptionem à
Millero nostro mihi traditam, anglicè reddendam, et in R. Societatis
consessu legendam curavi, magno cum audientium applausu. Grates
Tibi de illa decrevit cœtus, quod Tibi mandatum significandi provinciâ
in me lubens accepi. Serius quidem hæc perpetrata sunt, eò quod inter
autumnales Societatis inducias ad manus pervenerit Æstri descriptio:
induciæ enim istæ labente utplarimùm junio, vel erumpente julio
incipiunt, desinunt verò sub octobris finem.

Quascunque

The wealthy merchant George Clifford (1685–1760) created a herbarium with more than 3,000 species. Scientific and decorative elements are combined in this image to enhance the herbarium's function as an object of status.

Opposite page:
Carl Linnaeus, also known as Carl von Linné after his ennoblement, is still a towering figure in the field of natural history to this day and he remains Sweden's most famous scientific name of all time. This portrait is the work of Alexander Roslin.

Linnaeus's *Systema Naturae* (1735) became a landmark work in the understanding of how nature should be organised and divided.

with which his specimens had been handled, which compromised their condition; the difficulties of preserving herbarium specimens causes anxiety even today.

Others who contributed to the herbarium include the gardener George London, the horticultural expert Mary Somerset, Duchess of Beaufort, the Anglo-African Edward Bartar (his mother was probably a local woman, his father a servant of the Royal African Company), who occupied a powerful position on the Gold Coast – plant collecting had many associations with the slave trade – and the Reverend Adam Buddle, who gave his name to that prolific purple plant, the buddleia. John Ray was not the source of many specimens, but he did provide the means of classifying the herbarium through the weighty three volumes of his *Historia Plantarum*.

Sloane moved in the highest intellectual circles of his day. He was a friend of the philosopher John Locke, whom he persuaded to contribute articles on natural history, weather phenomena and a medical case to the *Philosophical Transactions of the Royal Society*, of which Sloane was a fellow. That was in the 1690s, when Locke was living in retirement in Essex, grateful to be sent 'news from the commonwealth of letters into a place where I seldom meet with anything beyond the observation of a scabby sheep or a lame horse.' While Sloane's insatiable appetite for collecting and interest in human and natural oddities may seem to lack discrimination, his aim was to form a

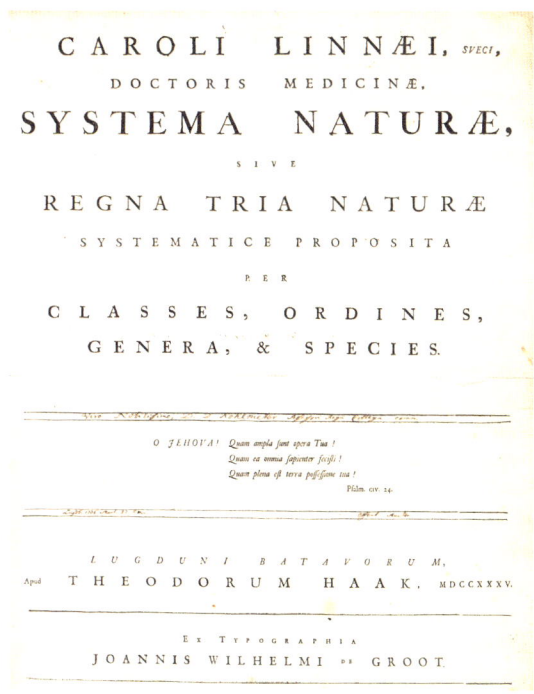

comprehensive collection in which everything in the world was represented. While the scale of his achievement was vast, he was not alone in his tastes, which were not so very different from those of Rudolf II.[4]

Sloane's interests, and those of the Royal Society, were shared by enquiring minds around Europe. The Sherardian Professor of Botany at Oxford was a German, Johann Jacob Dillenius, who had originally come to England to help the botanist William Sherard organise his herbarium (Sherard, in further signs of internationalism, studied for five years at the Jardin du Roi in Paris and edited the German botanist Paul Hermann's book on the botanical garden at Leyden University; he made a fortune as British Consul in Smyrna for the Levant Company). Sherard insisted that Dillenius should be the first person to hold the chair of botany that he endowed in his will.

In 1736, the great Swedish botanist Carl von Linné, otherwise known as Linnaeus, came to England, visiting both Sloane and Dillenius. Early in his career, Linnaeus had given a copy of Sloane's book on the natural history of Jamaica to George Clifford, a director of the Dutch East India Company, who had a botanical garden on his estate at Hartekamp, near Haarlem, in Holland; Linnaeus became his curator and described the garden in *Hortus Cliffortianus* (1737). It was Clifford who paid for the trip.

Linnaeus was born in Råshult, in southern Sweden, in 1707, the son of a minister. Unlike Sloane, who became a rich man through medicine and marriage, Linnaeus followed the passion for botany that he had acquired as a boy, becoming Professor of Medicine and later rector at Uppsala University, as well as a physician and eventually a royal physician to the admiralty. At the time, however, medicine and botany were closely connected, and it was in the medical faculty of Uppsala University that he found teachers. Sloane was an empiricist, more interested in the effectiveness of medical treatments than any theory about them. By contrast, Linnaeus wanted to go beyond mere observation and find the principles on which the natural world was organised; in 1735, he set out his ideas in *Systema Naturae*, stating that:

> The first step in wisdom is to know the things themselves. This consists in having a true idea of the objects; objects are distinguished and known by classifying them methodically and giving them appropriate names. Therefore classification and name-giving will be the foundation of our science.

One of Linnaeus's lasting contributions to science was the binomial system. The previous system had been cumbersome. Linnaeus devised a means by which every living thing could be defined in terms of two Latinised words, with a genus giving the type and the species a description – thus, most famously, *Homo sapiens: homo* means human being, while *sapiens* identifies a salient characteristic. However, his hope of creating a system that embraced every living thing in an exhaustive taxonomy was, we now realise, not realistic; nature changes. It defies rigid classification.

Contemporaries had another objection. Linnaeus based his classification of plants on the stamens and pistils of their reproductive mechanism, rejoicing in the promiscuity of bees as they wandered indiscriminately from plant to plant, taking nectar where they would. He courted controversy by using terms such as *monandria* (one husband) and *diantria* (two husbands), and the Prussian botanist J G Siegesbeck was duly outraged. 'What man', he railed in 1737, 'will ever believe that God Almighty should have introduced such confusion, or rather such shameful whoredom, for the propagation of the reign of plants. Who will instruct young students in such a voluptuous system without scandal?'

Linnaeus acknowledged that his sexual system was artificial. But he was also a pragmatist, and the system was undoubtedly very easy to use. We now know that his method did not always order nature into the correct families. But he was right in his belief that affinities between species did not depend on mere resemblance – a view to which Darwin refers in *On the Origin of Species*. Once it was generally accepted, Linnaeus's method proved remarkably robust, the principles on which it was based not being challenged or dismantled until the emergence of Darwin and, much later, the discovery of DNA in the 1950s. Yet, Linnaeus's system lives on in large part to this day.

Naturally, Linnaeus's herbarium, which is now kept in a fireproof vault beneath a wing of Burlington House, on Piccadilly, in London, was arranged quite differently from Sloane's. Sloane had the paper sheets of his herbarium bound into heavy books – reflecting, perhaps, the literary basis of education in the seventeenth and eighteenth centuries. Once bound, the sheets were difficult to reorder. Linnaeus kept his sheets loose, something he learnt from Clifford. This became the standard method for subsequent herbaria, including that of the Timm family. Not unlike the Timms, he ordered a number of rough wooden cabinets to house his collection. Wholly different from the sumptuous furniture such as the Badminton Cabinet, commissioned in Florence for Badminton House, in which aristocratic collections were kept, they were divided into small compartments ('pigeonholes', as they are popularly known) where the sheets could be placed. The Linnaeus Herbarium was a working collection. Although he sometimes acquired paper intended for other collectors and decorated with engraved vases out of which the specimens could appear to grow, he paid no attention to these ornaments. Some were brutally truncated. For the most part, he used what appear to be folio pages cut into two. Once ready, the sheets were filed in the pigeonholes. Each pigeonhole represented one of Linnaeus's 24 classes; inside it were folders containing every example that Linnaeus could find of the species belonging to it. It was a flexible system that could be expanded as new specimens were added – a three-dimensional equivalent of the index cards Linnaeus used to organise his notes.

Fish were preserved in exactly the same manner as plants, being cut in half and then squeezed under a weight to expel the moisture. In all, there were over 14,000 sheets. And, like other collectors, Linnaeus did not restrict himself to plants; there are also drawers of moths, butterflies and vividly coloured beetles – more than 3,000 insects and over 1,500 shells. Whereas the chlorophyll in plants turns brown over time, the iridescence of beetles' wings is just as spectacular today as it was in the eighteenth century.

Some historians have come to interpret Linnaeus's taxonomy as a means of facilitating the exchange of information around the world, as more of it was discovered by Europeans. Empire is never far away, according to this mindset. Nor were the practical advantages that could accrue from the study of new plants and plant relationships. Linnaeus hoped to improve Sweden's economy by reducing the money being spent overseas on luxury goods such as coffee, tobacco, tea and silk. In this he failed; only small amounts of silk could be produced from the white mulberries that a student of Linnaeus grew in the 1760s, and the silkworms that fed on them refused to make do with nettles, despite, according to Linnaeus, a relationship between plant and tree. But much was learnt about mulberries in the attempt, and Linnaeus had better luck with cultured pearls. Drilling a small hole in the shells of some mussels, he inserted a particle of limestone; the mussels were returned to the riverbed and half a dozen years later were found to have produced a spherical pearl – the world's first artificially created spherical pearls. (The technique was based on the misconception that pearls in oysters develop from grains of sand, rather than, as is now thought, a parasite or infection; but it nevertheless worked.) It was for this that Linnaeus was ennobled by King Adolf Frederick.

Naturally, Linnaeus valued his herbarium highly and left instructions in his will about its future. After his death in 1778, it was secure in the hands of his son, Carl Linnaeus the Younger, who was also a botanist. But the latter, dying in his early forties, only outlasted his father by five years, leaving his widow in need of support and without any professional interest in the collection. Such was Linnaeus's prestige – and fashionable interest in botany – that it was bought by the English amateur botanist Sir James Smith for a thousand guineas; the purchase gave Smith an instant standing among scientists. In 1788, he founded the Linnean Society, which in turn went on to found other learned institutions, such as the Zoological Society of London. On his death in 1829, he left both Linnaeus's herbarium and his library, which he had also purchased, to the society, where it remains. Smith's own herbarium of over 27,000 specimens survives in its original state, down to the mounts and labels.

Although the Timm Herbarium was not begun until the next century, a connection with Linnaeus can be made through Carl Peter Thunberg, one of Linnaeus's students; Gabriel Casper Timm had attended his lectures at Uppsala.

Linnaeus rarely travelled beyond Sweden and never left Europe, relying on his friends and disciples in the New World to search for useful plants on his behalf. But other botanists were more adventurous, tempted to undertake the dangers of exploration in order to see and record the wonders that were being opened up for the first time to European eyes. The French botanical explorer Jean Baptiste Fusée-Aublet risked snakes, sinkholes, fugitive slaves, disease, mosquitoes and a dreadful climate to assemble the large herbarium of the plants he discovered in French Guinea. Surely, he wrote, the botanists who endured such things for the sake of an uncertain reward could not have been motivated merely by self-interest? Admittedly plant discoveries were eagerly studied for their economic potential, but that came long after the hunt through previously uncharted territories had taken place. Botany, he wrote, attracted a particular type of individual: physically strong, determined, passionate,

light-hearted, discriminating and possessed of senses that were finely attuned to the nature around him.[5] Such a one was Fusée-Aublet, of course; the description also applies perfectly to Joseph Banks, whose extraordinary herbarium survives, like the Sloane Herbarium, in the Natural History Museum. Since Banks purchased Fusée-Aublet's herbarium, these two great collections are now one – although a small set of Fusée-Aublet's specimens also found their way into the herbarium of the philosopher Jean-Jacques Rousseau.

While Linnaeus was a professor, in need of a salary, Banks, as the son of a wealthy Lincolnshire landowner from whom he inherited as a teenager, had independent means. Educated at Harrow, Eton and Oxford, he did not excel at the Classics which constituted most of the educational regime; but he developed a passion for botany, largely outside the formal curriculum.

When he came into his fortune at the age of 21, he left Oxford for London. There, he became a fellow both of the Royal Society and the Society of Antiquaries – natural history and antiquarianism would remain bedfellows into the nineteenth century – while, at the British Museum, he made a friend of the Swedish naturalist Daniel Solander, who was the assistant librarian and had trained under Linnaeus. Solander formed one of the party that Banks took with him, at his own expense, on Captain Cook's voyage of exploration on board HMS *Endeavour*, which left England in 1768, not to return for three years. Banks's secretary was a Finn, Herman Spöring, and there were also two draughtsmen, Alexander Buchan and Sydney Parkinson. While Parkinson, who was principally responsible for recording the specimens of natural history, could draw from nature, if a plant was brought back in fit condition, it was often more practical for Banks, who might be away from the ship for many days, to press it between leaves of paper. To this end he brought a supply of old newspapers, some of them already 30 years old. Pressing and drying were essential to keep the plants that Banks gathered in a condition to survive the long journey home.

Cook's expedition was not the only one to set sail from Europe. It followed in the wake of Louis-Antoine, Comte de Bougainville, who, earlier in the 1760s, had circumnavigated the globe – the first Frenchman to do so.

Linnaeus himself had recommended that Philibert Commerson should be taken by Bougainville as a botanist. Commerson travelled with a delicate valet, who turned out to be a woman called Jeanne Baret; although sailing under false colours, she has the credit of being the first woman to complete a circumnavigation. They stopped at Tahiti, explored Mauritius and (after Bougainville's return to France) studied the flora and fauna of Madagascar. Commerson named the Bougainvillea, collected in Brazil, after his captain, whose name also attaches to Bougainville Island, part of Papua New Guinea. Cook's mission took him to the great southern continent, whose existence had been supposed but not proven. Having observed the transit of Venus, he made running surveys of the coasts of New Zealand and Australia.

Some of the greatest scientific names of the eighteenth century are found in this painting from 1771 by John Hamilton Mortimer. From the left, Linnaeus's disciple Daniel Solander and the botanist and scientist Sir Joseph Banks, both of whom followed the explorer Captain James Cook (centre) on his travels. To the right of Cook is the author and editor John Hawkesworth, who was responsible for the publication of Cook's accounts of his voyages. On the far right is John Montagu, the Earl of Sandwich, who financed the expeditions.

A six-week stop for repairs in the Endeavour River enabled Banks to botanise inland, collecting herbarium specimens as he did so. In Tahiti, Banks wrote the first European account of tattooing. Illnesses contracted in Batavia led to the death of a third of the crew on the *Endeavour*, along with Buchan and Parkinson from Banks's party. Like Cook, Banks survived, returning to England in 1771.

Well-connected and dashing, Banks became feted. Fortunately, he was also by nature a scientist who used his position, knowledge and organising ability to enormous effect. George III, the Farmer King who numbered botany among his many interests, appointed him the informal director of the botanical garden he had established at Kew, site of a royal palace west of London. In expanding Kew, Banks was able to call upon the many scientific contacts he had made around the world. The King's name could be invoked to encourage ambassadors and the East India Company to help in collecting plants and seeds. Banks was a systematic man. Hoping to rival and surpass similar institutions in Paris and Vienna, he organised a programme of plant collecting that would be as comprehensive as possible. Gardeners from Kew were sent to South Africa, the Canaries, Australia, South America and elsewhere. They not only sent back plants, but details of the soil they grew in and the conditions in which they flourished. Two gardeners from Kew were on HMS *Bounty* when Captain Bligh sailed to Tahiti to collect breadfruit plants, to be raised and transplanted as a source of cheap food in the West Indies. Banks's instructions for the care of the plants on the cramped ship, which involved frequently moving and sponging them, may have contributed to the famous mutiny.

Kew Gardens may serve both as a site of scientific research and a pleasure ground for Londoners, but George III's hope was that its botanical work would improve British farming. Moreover, the exchange of plants from around the world would elevate the condition of humanity generally. In a letter of 1787, Banks expresses this ambition in rousing terms, in connection with the botanical garden that had been proposed for Calcutta:

To exchange between the East & West Indies the productions of nature usefull for the support of mankind that are at present confined to one or the other of them, to increase by adding this variety the real Quantity of the produce of both Countrys, & by that means their population, furnishing at the same time to the inhabitants new resources against the dreadfull effects of Hurricanes & droughts, to one or the other of which all intertropical countries are subject, are the more immediate objects of his Majesty's present intentions, the disinterested humanity of which seems alone sufficient to inspire diligence & activity into the minds of all those who may be fortunate enough to be allowed a share in the honor necessarily consequent in having carried ideas of such exalted benevolence into Execution. [sic]

The Botanical Magazine was founded by the botanist and entomologist William Curtis (1746–99). First published in 1787, it still exists today and is considered the world's oldest botanical journal. This image is a hand-coloured illustration of *Artocarpus incisa*, the breadfruit tree that comes from Southeast Asia.

Opposite page: Joseph Banks's herbarium as it looked in 1820, located in his London home. The room rises through two floors and also has skylights, which provided ideal conditions for studying the collection. It was environments like this that inspired other collectors to display their natural history collections in a modern and scientific way.

Science and altruism combined with Britain's desire to challenge France and Holland over the command of the lucrative spice trade: imperial self-interest happily coincided with the Enlightenment ideal of bettering the human condition. Herbaria were central to the project.

The practical value of dried plant material was clearly in the mind of James Smith, the purchaser of Linnaeus's herbarium, since he assembled a collection of economic botany – 900 examples of seeds, fruits, wood, fibres, fungi, whale skin and the wing of a flying fish. This displays the usefulness of botanical knowledge, as found in the discoveries being made in distant lands. In this he was typical of an age that was captivated by the romance of exotic and far-away places, but also hard-headed about the benefits that could accrue from possessing them and their treasures.

To some, dried plants cannot have been as exciting as other wonders that Banks brought back from his voyage, such as the previously unseen skin of a kangaroo, which would have been, until then, unimaginable. Such extraordinary objects fed an appetite for the natural world, among a public already excited by the debate on the new way of viewing landscape, called 'the picturesque'; gentlemen tried to make the parks around their houses look like paintings by Claude Lorrain. Edmund Burke introduced the concept of 'the sublime': fearsome manifestations of nature could be pleasurable, as long as the viewer did not feel physically threatened – a doctrine that applied to the vastness of the landscapes encountered by explorers. Those who could afford it created grottoes, inviting visitors to wonder at the strangeness of nature in a setting of trickling water and shadows. They were decorated with minerals and shells in artistic designs.

The poet Alexander Pope described his famous grotto at Twickenham, beside the Thames, in couplets. Reflections from the river shone like

> …a broad Mirrour thro' the shadowy Cave;
> Where lingering Drops from Mineral Roofs distill,
> And pointed Crystals break the sparkling Rill.

Geological specimens also feature in the grotto, created on a larger scale, by Alexander Hamilton, at Painshill Park, in Surrey. The ceiling is a mass of stalactites made from plaster, to which wafers of shiny, sparkling felspar have been attached. Among the felspar are lumps of amethyst and other crystals. A gardener waited in the dark for visitors to approach, then switched on a pump to circulate dripping water – while the surface of a pool remained still, to produce reflections. Presumably Hamilton's intention, like Pope's, was to inspire wonder at the natural world, in contrast to the introspection of Painshill's hermitage or the gloom of the Ruined Abbey (erected to hide a brickworks). Ladies such as the Duchess of Richmond were attracted to colourful shells; she and her daughters took seven years to decorate the grotto at Goodwood House in Sussex with an intricate pattern of shells enlivened with mirror glass. It was a rich person's hobby, due to the exorbitant cost of exotic shells. Officers in the Royal Navy were tasked to bring back colourful examples from their cruises. Ladies without such exalted means or connections, such as the Misses Parminter at A-la-Ronde, their *cottage orné* on the Devon coast, gathered shells from their travels or the seashore. The patterns that they made from them could be a triumph of handicraft.

During the next century, ever more fanciful ways were found of pressing nature into the service of decoration. Furniture made from antlers, terrariums filled with mosses, ostrich feathers on hats, floors patterned with the knuckle-bones of deer – all had their day. Closer to the herbarium aesthetic was the rage for preserving seaweed, in all its variety of silhouette. A B Hervey's

The French artist Alexandre-Isidore Leroy de Barde (1777–1828) continued the long tradition of still life drawings and paintings of nature. Here is his depiction of crystalised minerals arranged in different compartments. Capturing their many colours and textures placed high demands on an artist's ability. In this example, the result is almost photographic.

Sea Mosses: a collector's guide and an introduction to the study of marine algae, published in Boston in 1881, explains the process by which it was dried and mounted:

> The tools needed are a pair of pliers, scissors, a stick with a needle in the end, at least two 'wash bowls', botanist's 'drying paper', or some kind of blotting paper, cotton cloth, and finally cards to mount the specimens on. Pliers and scissors are used to handle the specimens and cut away any extraneous, 'superfluous' branches, and the needle is used like a pencil so that the plant can be moved around with relative ease to show the finer details... The drying and pressing process consists of layering the mounting papers with various types of blotting cloth and additional paper topped with weights... Most seaweed in this case will adhere to the mounting board via gelatinous materials emitted from the plant itself.

The result would have appealed both to inquisitive and artistic personalities.

The Timms were less exuberant in their displays. Boxes of birds' eggs and insects are arranged without fanfare, in the soberest of trays. But clearly the collecting of natural objects had entered the mainstream of nineteenth century life. It was magnified by a boom in horticulture. Exchanging plant material had been an activity of the rich in the old Europe. There are records of packets and collections of seeds having been sent out from Coutts Bank, where Banks later kept his account, as early as the 1740s; the recipients must either have been friends of the Coutts family or clients of the bank.

Opposite page: The market for seeds grew dramatically during the nineteenth century as interest in horticulture increased. Catalogues like this were widespread and often had richly decorated front pages that stimulated the interest in acquiring seeds.

A group of Victorian gentlemen standing outdoors photographed during the second half of the nineteenth century. Collecting natural history objects was a very popular social activity.

SUTTON'S

AMATEURS GUIDE

AND

SPRING CATALOGUE

FOR

1865

CONTAINING
MUCH USEFUL INSTRUCTION IN

HORTICULTURE & AGRICULTURE

SEEDSMEN TO THE QUEEN
AND
H.R.H THE PRINCE OF WALES

SUTTON & SONS

Royal Berks Seed Establishment,
READING.

James Coutts bought from Minier and Mason, seedsmen on the Strand, who opened an account in 1757. But their activities were modest in scope beside those of the great Victorian nurserymen and seed merchants, able to benefit from the growth of a horticulturally dedicated middle class and the burgeoning numbers of country house owners, who wanted to grow the latest plant discoveries from around the world. So big did some companies become that they financed their own plant-hunting expeditions, to bring back new species that could be adapted for English gardens. For example, the Veitch family, operating from Exeter and the Royal Exotic Nursery in London, sponsored Thomas Lobb to find orchids in Asia, and his brother, William, to scour the Americas – he brought back the Wellingtonia tree (*Sequoiadendron giganteum*) and the monkey puzzle tree (*Araucaria araucana*). By 1914, Veitch Nurseries had introduced nearly 1,300 plants into cultivation. Other plant hunters were sent by the Horticultural Society of London (known as the Royal Horticultural Society from 1861).

Horticulture provided a career structure for ambitious young men (professional gardeners were largely men until the advent of lady designers like Gertrude Jekyll at the end of the century), who had had a role model in Sir Joseph Paxton – the garden boy who became an engineer, the designer of the great conservatory at Chatsworth House, for the Duke of Devonshire, and the Crystal Palace, for the Great Exhibition, for which he was knighted. He was also an MP and a railway director.

It was not only the size of the market for botany that grew, but the number and types of people entering it. Weavers studied Linnaeus at their looms, botanised the fields around them and submitted the results to the great Sir William Hooker at Kew. By 1873, there was a sufficient quantity of amateur botanists of note for James Cash to write a book about them, with a title that evoked the self-help theories of Samuel Smiles: *Where There's a Will There's a Way! or, Science in the Cottage. An account of the labours of naturalists in humble life.* One of his subjects, John Horsefield, a weaver who had a single year's education before he started work at the age of six, had a daffodil named after him – *Narcissus horsfieldii*. Horsefield founded one botanical society and became president of another, in the Manchester Area; these working men's societies often met in public houses which provided rooms where herbaria could be stored. They were part of a network of natural history and antiquarian societies, which had covered the whole of the British Isles by the end of the century. The Somerset Archaeological and Natural History Society was typical, having been founded by 'several gentlemen of Taunton and its neighbourhood' in 1849. Two years later, it could boast 420 members, consisting largely of gentry and clergy. From 1874, they were quartered in Taunton's medieval castle, which they had purchased; here, they kept a herbarium, which was enriched by collections bequeathed by members at their death. Societies like this Somerset example existed throughout Europe and in parts

A caricature of the Scottish geologist James Hutton (1726–97) who 'sees' the 'faces' of stones. With his hammer raised, he is ready to hack out layers of exciting fragments to take home for his collection. Hutton was more or less self-taught and went on study trips around the UK where he visited many different types of landscapes of geological interest. He is considered to be one of the founders of modern geology.

of the United States. They provided an outlet for the botanical passions of people who had not been able to attend university, usually but not always for lack of money (in Britain, the founder of the Linnaen Society came from a Dissenting family, his father having been a Unitarian wool merchant in Norwich; universities were only open to members of the Church of England). The collecting of plants, birds' eggs and insects became a recognised hobby. When Arthur Conan Doyle made the villain of *The Hound of the Baskervilles,* published in 1902, an amateur entomologist, who carried 'a tin box for botanical specimens hung over his shoulder and… a green butterfly-net in one of his hands', he was evoking a familiar type.

A new urgency was given to the study of natural history by the theory of natural selection advanced by Charles Darwin. Linnaeus had assumed that the animal kingdom, in all its wonder and variety, had, like all living things, been put there by God, and he made it his task to categorise it; taxonomy would give an insight into the mind of the Creator. Although Darwin did not challenge Linnaeus's system of classification, he introduced the idea of natural selection, which later came to affect our view of the world, human beings and nature. To the student of evolution, certain types of butterflies and moths are particularly interesting because of the speed at which they can adapt, changing colour according to the backgrounds against which they must camouflage themselves. The peppered moth, for example, began the nineteenth century as a predominantly white insect, with a black speckle to its wings; but in the soot-encrusted cities of the industrial revolution, it turned black. Were Gabriel Casper and Paul August Timm aware of such phenomena when they collected insects at Engelsberg? At the least, they would have known of the controversy that surrounded Darwin's work. Perhaps they were driven by the desire of all collectors to achieve a feeling of completeness, by acquiring examples of every insect that fitted their collecting criteria. We can only speculate on the discussions they may have had as they examined the serried ranks of their specimens, carefully pinned and labelled in the many drawers of their cabinet.

Rocks were even more challenging to the religious. In the late eighteenth century, the Scottish geologist James Hutton had deduced that the different layers of rock that were visible where the sea had stripped them bare of later accretions were the result of tremendous forces acting upon the earth's crust; this was set forth in his *Theory of the Earth.* Hutton's biographer John Playfair recalled the excitement of the moment as Hutton propounded his ideas on a headland near Berwick-upon-Tweed to which they had specially sailed: 'What clearer evidence could we have had of the different formation of these rocks, and of the long interval which separated their formation, had we actually seen them emerging from the bosom of the deep… The mind seemed to grow giddy by looking so far into the abyss of time.' Although Hutton emphasised his belief that the processes by which the earth had been shaped were the work of a beneficent Creator, his observations contradicted the Biblical account of the origins of the world.

Even greater anxiety was caused by analysis of the fossil record. Fossil collecting was a Victorian passion, combining a healthful scramble over rocks with intellectual improvement. The aesthete and seer John Ruskin, always hypersensitive to the natural world, was a keen amateur geologist; from boyhood, he had loved breaking open stones to discover the crystals hidden inside. But the discovery and interpretation of fossils by professional scientists challenged the belief system in which he had grown up. As he wrote in a letter, 'If only the Geologists would let me alone, I could do very well, but those dreadful Hammers! I hear the clink of them at the end of every cadence of the Bible verses.' As iron founders, the Timms took a special interest in geology, from the point of view of minerals rather than fossils; the specimens in their collection may have served as a working library. Their close involvement must have made them aware of contemporary debates surrounding this branch of science.

Just as every seventeenth-century princeling had a cabinet of curiosities, so every imperial capital of the Victorian period had a herbarium, as did many industrialists and inquisitive gentlemen.

One of the biggest in the world was, and still is, at Kew, where William Hooker was appointed director in 1841. To begin with, Hooker's very large collection was run, as Banks's had been, on a gentlemanly basis, from his own home close to Kew. But eventually new premises were found, a curator was appointed, and other collections were added to Hooker's foundation; these included the herbarium of the East India Company and some specimens from the Petiver Herbarium, collected in 1696. The Kew Herbarium now contains specimens from 95% of all the vascular plants in the world and continues to grow at a rate of 25,000 additions a year. While Kew represented an advance in how to organise a public, scientific institution, private herbaria remained an essential interest of the new ranks of millionaire industrialists and businessmen. Some assembled herbaria on a professional scale, conceived as contributions to science and ultimately bequeathed to universities. Henry Borron Fielding, whose father had owned a business printing calico in Lancashire, left his large collection to Oxford University, where it was later combined with that of George Claridge Druce, a retired pharmacist; this formed the Fielding-Druce, which incorporates those of Sherard and Dillenius among others. While the Timms had more modest ambitions for their herbarium, they were not alone as independently wealthy collectors.

Herbaria, however important and extensive, are vulnerable to poor curatorship and neglect. Even the famous collection of Meriwether Lewis and William Clark, made while crossing the west of the United States for the Corps of Discovery Expedition that followed the Louisiana Purchase of 1804, underwent numerous vicissitudes before the remaining 239 sheets were safely lodged with the Academy of Natural Sciences in Philadelphia.

With today's photographic techniques, we are able to come far closer to the details of plants than the Timms ever could. The patterns, colours and textures that are revealed by modern magnification inspires the same kind of awe that the use of a loupe or magnifying glass will have given to those studying plants in the nineteenth century.

They are now studied as evidence of the state of plant life before the Industrial Revolution. For this is the remarkable truth about herbaria of whatever size and date around the world: they continue to be of value to science, however well known the species to be found in them may be.

It is herbaria that contain the botanical types of plant species, meaning the examples that provide the benchmark for identification. These are often examined for DNA samples and other research purposes, and are a vital part of today's science. The seed banks of our time store seeds from the world's plants, dried to a moisture content of around 5% and frozen to minus 20 degrees Celsius; in this form, they can last for hundreds if not thousands of years. However, after being gathered in the wild, it is essential that they are correctly identified. This is done by cross-referencing the seed-bearing specimen with the type in the herbarium. Without herbaria, seed banks, such as the Royal Horticultural Society's Millennium Seed Bank at Wakehurst Place in West Sussex, which stores 2.4 billion seeds from around the world in underground vaults, would be impossible.

Ordinary herbarium specimens can also be useful; their labels show exactly where and when they were collected, enabling scientists to understand the spread of plant populations and how they respond to the seasons. To quote a recent article from the *Philosophical Transactions of the Royal Society*, 'Herbarium specimens provide verifiable and citable evidence of the occurrence of particular plants at particular points in space and time, and are vital resources for assessing extinction risk in the tropics, where plant diversity and threats to plants are greatest.'[6] And a journal of plant science notes 'a growing trend in which scientists are taking advantage of herbarium and other museum specimens to study ecological responses to global change. Herbarium specimens allow researchers to gather phenological data covering broad geographic and temporal ranges, in a way that is not possible using almost any other method.'[7]

And so herbaria, whose contents are both dry and dead, are contributing to lively, young methods of science. Collections such as that formed by the Timm family are now digitised and shared with a far larger public than could ever view the fragile herbarium sheets. Imaginative photography, such as that commissioned by the Ax:son Johnson Foundation, provides new insights into the beauties of the vegetable kingdom, revealing the architecture of nature to the eyes of the twenty-first century.

1. Barbara M Thiers, *Herbarium: the quest to preserve and classify the world's plants* (Portland, Oregon, 2020).
2. Thomas Sprat, *The History of the Royal Society of London, for the Improving of Natural Knowledge* (London, 1667), pp. 72–73.
3. Mark Carine (ed), *The Collectors: creating Hans Sloane's extraordinary herbarium,* (London, 2020). A catalogue exists in James Dandy, *The Sloane Herbarium: an annotated list of the Horti sicci composing it; with biographical details of the principal contributors,* (London, 1958).
4. See 'Sir Hans Sloane, Scientist', by Maarten Ultee, in *The British Library Journal,* Spring 1988, Vol. 14. No 1, pp. 1– 20.
5. Jean Baptiste Fusée-Aublet, *Histoire des plantes de la Guiane Françoise* (London and Paris, 1775).
6. Nic Lughadha E et al, 'The use and misuse of herbarium specimens in evaluating plant extinction risks' (2018), in *Phil. Trans. R. Soc.* B 374: 20170402. https://doi.org/10.1098/rstb.2017.0402
7. Richard B Primack and Amanda S Gallinat, 'Insights into grass phenology from herbarium specimens', in *The New Phytologist,* March 2017, Vol. 213, No. 4, pp. 1567–68.

One of the most colourful
natural history collectors
of the eighteenth century
was the extremely
wealthy Duchess of
Portland, Margaret
Cavendish Bentinck
(1715–85). She created
what was said to be the
largest collection of all
time. She hired a team of
botanists, entomologists,
ornithologists and others
who arranged and
developed the collection,
among them Daniel
Solander. After her death,
the collection was sold
and the objects dispersed
through an auction that
took weeks to complete.

A cabinet of curiosities in a simple wooden and glass structure with corals, paintings, graphics and other remarkable things produced by Domenico Remps (1620–99)

Opposite page: A masterly still life with corals, shells and plants from the bottom of the sea, depicted by Anne Vallayer-Coster (1744–1818).

Page 44–45: A playful still life with a garland, masks and bows, formed by shimmering seashells of different colours by Jan van Kessel the Elder (1626–79).

Page 46–47: In Amsterdam, the wealthy textile merchant and designer Levinus Vincent and his wife Joanna created an admired natural history collection. A catalogue was also printed and it is said that they were inspired by the beautiful patterns of their fabrics in the creative arrangement of the natural objects.

Another of Joris Hoefnagel's imaginative still lifes. The insects are represented in such detail that it feels as though they are floating towards us with the help of a breeze blowing from the mouth of Aeolus, the Keeper of the Winds in Greek mythology, who is visible in the background.

Despite the fact that the collecting of natural objects was conducted on different scales, by different people and in different countries, there are startling similarities in how they were presented. In this painting from 1785 by the British artist Sarah Stone (1760–1844), Sir Ashton Lever's Museum of Natural History display included items found by James Cook on his voyages to Australia and the Pacific.

Following pages: Still life by Frans Francken the Younger (1581–1642). In the section to the right, there is a glimpse of two men, perhaps in conversation about one of the interesting objects on display.

The Timm family natural history collection at Engelsberg Ironworks

SVANTE HELMBAEK TIRÉN

… the splendour of the flowers, for which he held so much love, to which he eagerly devoted so much research, and of which he acquired such deep knowledge. Possessing a more than average memory – as well as being a faithful and diligent practitioner of the arts and sciences, he was even well versed in other areas of human knowledge (learning), although he was not one to boast about his abilities. Humble, modest, unassuming in this, as in all else, he resembled our mountains of ore, that hide their rich treasures within their depths, until the day an industrious hand brings them to light.

Taken from 'Words spoken about the ironworks proprietor Gabriel Casper Timm at his funeral at Engelsberg…' by P A Ljungberg, 1871.

The Timm family natural history collection, resting in its sturdy wooden cabinets, is a remarkable record which harks back to a time when a whole world of ideas could be expressed through the act of collecting. Nothing is revealed from the outside of the cabinet, there are no signs or identifying markings of any kind. But inside reside thousands of flowers, herbs, grasses and other plants, painstakingly mounted on thin sheets of paper.

With care and respect for this delicate and fragile material, the visitor can carefully pull out one of the drawers, gently lift off the tissue paper, and dive into a world of dried flowers. Both wild and cultivated species are to be found, some magnificent, others inconspicuous, the common and the rare. Some originate from places scattered around the Nordic region, while others are picked at home, in the Västmanland countryside surrounding Engelsberg Ironworks. A number of them come from the garden itself, which is still maintained on site to this day. Among other things found in the herbarium are majorwort (*Astrantia major*), greater quaking-grass (*Briza maxima*), mallow (*Malva verticillata* var. *crispa*) and daisy (*Bellis perennis*), which once grew in the estate's park.

When exploring the collection, one can't help but feel the sense of wonder which must have been shared by its creators, Engelsberg's proprietor, Gabriel

'Engelsberg bruksegendom' (Engelsberg estate) depicted in the weekly magazine *Ny illustrerad tidning* in 1866. Beyond the trees lies a glimpse of the white manor house where the natural history collection was stored. Illustration by Johann Friedrich Meyer.

POLEMONIUM COERULEUM Linn.
Ad rivulum prope diversorium Upland Jemtiæ.
Aug. 1844: C. Lagerheim et G. Sjögren.

Casper Timm (1799–1870), and his son, Paul August (1831–78). Their passion for collecting is evident throughout the collection, a great tribute to their achievements. Most of the plants are captioned by the name of the species, its discovery site, collection date and the finder's name, all noted neatly by an elegant hand or on a printed label. Sometimes it is the father, sometimes the son who discovers an interesting new species while strolling around; at other times, it materialised from one of the many other collectors and scientists they knew. After pressing, the collected plant was attached to the paper with the help of small thin strips of paper that held the flower in place, without obscuring it from view. The plants are of course impressive, but so too is the attention to detail and the great patience that must have been required to put the collection together. The Timms were amateurs in the best and original sense of that word. They felt a strong connection to and love for their subject, even if it was not their professional undertaking.

It is almost uncanny how well-preserved the contents of the herbarium are, as is the rest of the natural history collection. For it was not botany alone that interested father and son. In other cabinets, we find drawers containing brilliantly coloured butterflies, beetles and other insects, shimmering stones and minerals, birds' eggs and seashells. Carl Linnaeus's classification of the world into three kingdoms was of course followed, those of plants, animals and minerals. The collection also included coins, scientific instruments, stuffed birds, books and other things indicative of an inquisitive scientific mind. But the natural history objects themselves lay at the core of the collection, and became the hub around which they built a private sphere beyond the chores and other aspects of everyday life.

Collections of this kind are not unusual. They can be found in different forms in both private and public ownership, for example on country estates, larger farms and palaces, or in university or museum collections. Some of these collections are only partly intact, while others, like the Timms', are more or less complete and untouched. In some collections, fragments of several collectors' work from different periods are mixed together. What makes the Timm family collection unusual is that it has been preserved in its original context and condition – even in its original cabinets – complete with documentation, accounts, literature and notes that allow parts of the collection to be examined in a way that isn't possible with other collections. It therefore provides us with insight into the fascinating world of private collecting in the nineteenth century.

Guided by the work of father and son Timm, we gain insight into the thinking and experiences that led to the creation of these types of collections. The material certainly warrants more study, especially as many of the plants and animal species in the collection are becoming increasingly rare today. At the same time, the study of the ideas that motivated the Timms to dedicate themselves to nature is a valuable beginning in the exploration of how a natural

Polemonium Caeruleum Linn belongs to the blue-gold genus (Polemonium) and was a popular garden plant. Note how small strips of paper were used to carefully secure the flower, which has been bent to fit on the page.

Lovisa Ulrika, along with her husband Adolf Frederick, was considered to be one of the foremost natural history collectors of her time. This portrait of her is by Lorens Pasch the Younger.

Opposite page: The Augsburg art cabinet is like a miniature universe, representing the entire world of thought of the seventeenth century. It was constructed under the direction of art dealer and diplomat Philipp Hainhofer (1578–1647) and given as a gift to Gustav II Adolph in 1632.

history collection can be regarded as part of Swedish cultural history and as having universal value.

There are many other collections of this kind, which also deserve greater attention and use, although it isn't always easy to see how this might be done. But it would certainly please the Timms to know that their collection had become part of a larger discussion about our natural world, and that it had helped us to understand and appreciate it better.

Natural History collection during the 1700s and 1800s

The Timms were not unusual in their fascination with natural history. Admiration for the natural world and the collection of its objects is a fairly basic human trait that appears in endless variations in different cultures over time. The ancient Roman work attributed to Pliny the Elder, *Naturalis Historia* ('The Natural History'), is often highlighted in the history of ideas of the West, as it contains early examples of, among other things, reflections on botany, zoology and mineralogy. During the Renaissance, a more systematic or museum-like collecting methodology began to take shape, not least in the form of the cabinets of curiosities or *Wunderkammers* described in more detail in the previous chapter. These became commonplace at royal courts and in the homes of the nobility and eminent scientists of wider Europe.

Objects and specimens were typically grouped in their own sections, occasionally given a primary focus, with some included in the collections of art objects and antiques. As early as the seventeenth century, in Sweden, a few cabinets of curiosities and nature chambers had begun to appear.

The well-preserved Augsburg art cabinet, which both Gabriel Casper and Paul August Timm must have seen during their studies at Uppsala University, is one such example. The cabinet was a gift to Gustav II Adolph following his triumphant entry into the city of Augsburg in 1632. It was said the world's foremost rarities were gathered there in a myriad of small compartments and drawers, containing everything from medals, relics and instruments, to the riches of nature in the form of minerals, shells, fossils, seeds, dried plants and stuffed animals. The cabinet is adorned by a small mountain of rare exotic minerals and corals upon which sits a large Seychelles nut encased in silver, which forms part of a drinking cup – this, in and of itself, is surely a sign of the superior standing then accorded to the wonders of nature.

It is not so surprising that the collecting of objects from the natural world really took off in a big way in Sweden during the eighteenth century. The many scientific breakthroughs, led by Linnaeus's *Systema Naturae* from 1735, afforded Sweden great self-confidence as a culture-bearing research nation, and spurred an interest in collecting among both private individuals and scholars at schools and universities. King Adolf Frederick and Queen Lovisa

Opposite page: Several rooms at Drottningholm Palace were designed especially for Lovisa Ulrika's collection of natural history objects and coins.

Adolph Ulric (1752–97) and Anna Johanna Grill (1753–1809) at Söderfors bruk. They were, like the Timm family, both ironworks owners and dedicated natural history collectors. Together they built up one of the more prominent collections of their time.

Ulrika were both devoted collectors and were very important as patrons and role models. Lovisa Ulrika had several rooms at Drottningholm Palace furnished for her natural history collections, while Adolf Frederick had his own space at Ulriksdal. Linnaeus himself was commissioned to organise and describe the collections. In the preface to his publication *Museum Regis Adolphi Frederici*, Linnaeus states: 'Natural-history collections are without a doubt one of the most chosen and righteous pleasures…that a human being can choose to take part in, for all inhabitants of the world and even the most distant parts of it who share man's vision and thought.' The connection of the natural to the spiritual realm was also emphasised: 'The earth is nothing more than a natural-history collection of the all-wise creator.' This view was shared by many, both in and outside Sweden. The trade in natural objects developed into an extensive international market that constantly provided collectors with new specimens. Sometimes shockingly high prices were paid for rare specimens.

As well as the expansion of universities during that period, more and more scientific institutions and societies were also founded, attracting both amateurs and professional researchers to their halls. *Kungliga Vetenskaps-Societeten i Uppsala* (the Royal Society of Sciences in Uppsala) was founded as

early as 1710 and was followed by *Kungliga Vetenskapsakademien* (the Royal Swedish Academy of Sciences) in 1739. In 1753, Queen Lovisa Ulrika established *Kungliga Vitterhetsakademien* (the Royal Swedish Academy of Letters, History and Antiquities). People often moved freely between the subjects, from natural history and botany, to rural economy and agriculture, the preservation of antiquities, coin collecting and so on. There were plenty of specialists in the various areas, but, in public discussion, the climate encouraged openness between different disciplines. The line between the professional and the amateur was not always clearly drawn.

This glorious period faded somewhat during the Gustavian era (1772–1809), when other intellectual ideals took over. Examples exist of the perceived scientific pedantry and dry rationalism of the time being mocked in favour of other schools of thought. In the poem '*Porträtterne*' ('The Portraits') from 1797, the poet Anna Maria Lenngren ridiculed 'the widely travelled president', who knew 'the names of flies in Greek and Latin', and whose principal scientific achievement was the donation of an angleworm to the Royal Swedish Academy of Sciences.

Natural history collectors could be both admired and ridiculed. One is reminded of the evident breach between the Enlightenment and Romantic eras, even though it was hardly as simple as it has been portrayed in retrospect. Gustav III was not himself a natural history collector of the sort his

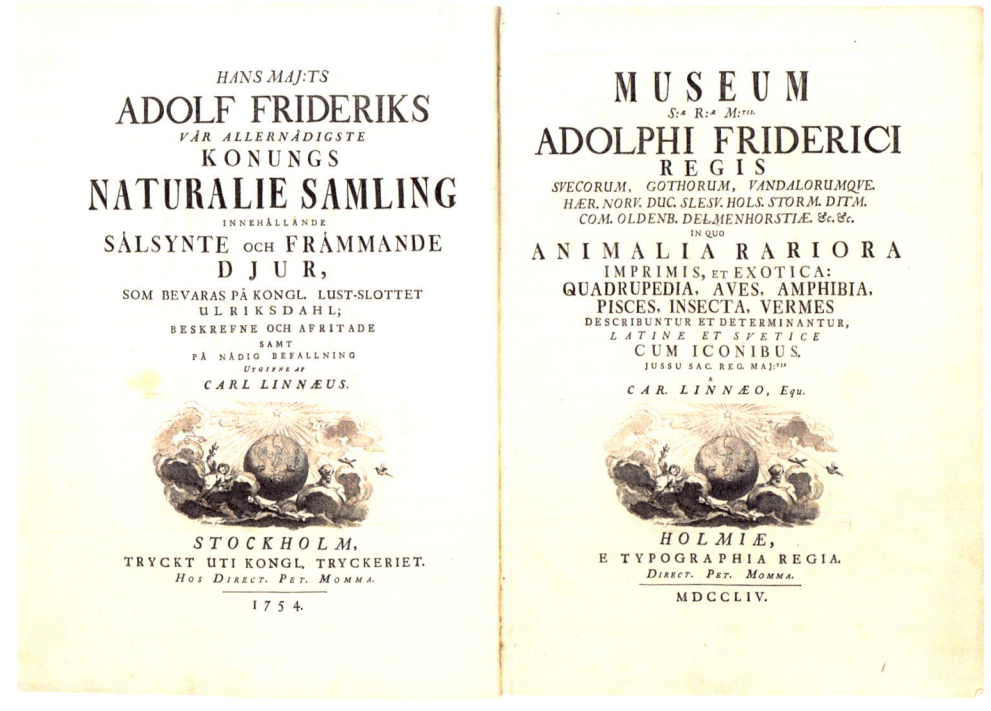

parents had been, even though the king most certainly had a great interest in science. But during his reign the fine arts and culture were given more emphasis. Yet, regardless of how natural history interests were portrayed in cultural discourse, they continued to be an important part of the life of intellectuals of the time. Scientific literature grew, and even newspapers, for example, constantly reported on its progress. By the end of the century, there was an impressive gallery of people who had natural history collections all over Sweden, who were in close contact with one another and kept up-to-date with news and discussions both at home and abroad. Linnaeus's grand plan for Sweden as a nation of science had been realised in some ways and would continue to contribute to a number of important events in the future.

Even though interest in natural history, like other types of collecting, went in waves and branched out in different paths, its development continued and spread during the late eighteenth and early nineteenth centuries. Across the country, estate owners, local mayors, assistant lecturers, sheriffs and priests were among many others who found themselves in the throes of collector fever. Collections were also being created at schools and educational facilities. Some collectors remained happy amateurs, while others hovered on the borderland with true professionalism, even though they lacked any substantial scientific education.

Even when one restricts the sample to estate owners and industrialists, a striking number of natural science collectors can be found. Some of these have gone down in history as legendary names that have contributed major donations or printed catalogues or studies that are still preserved today. Others have gone unnoticed among the substantial number of industrialists who had intellectual interests but no immediate need to be visible in that role. Many fundamental ideas were widely shared, but the level of ambition and how individuals regarded their own collection and its purpose varied.

Among the ironworks proprietors who were interested in natural history, the Grill family at Söderfors in Uppland can be mentioned as an important example. Adolph Ulric Grill (1752–97) and his wife Anna Johanna (1753–1809) built one of the most important collections of their time. They had the assistance of natural history expert and estate physician Pehr Lindroth, who over time served as a kind of curator for the collections. The couple and Lindroth travelled to England in 1788, among other places, where they exchanged a stuffed moose for a large number of exotic bird species.

After Adolph Ulric's death, Johanna continued to collect and even built a separate building for her natural history cabinet. In 1828, the entire collection was donated to the Royal Swedish Academy of Sciences, where it became the backbone to what would later become *Naturhistoriska riksmuseet* (the Museum of Natural History). It is not unreasonable to imagine that Gabriel Casper Timm was probably acquainted with the Grills, had possibly visited Söderfors before the collection was moved, and had certainly seen parts of the collection

The catalogue of Adolf Frederick's natural history collection was written by none other than Carl Linnaeus. The king's collection was kept at Ulriksdal Palace and later transferred into the possession of the Royal Swedish Academy of Sciences. It was later incorporated into what became the Swedish Museum of Natural History.

Owner and proprietor
Gabriel Casper
Timm (1799–1870)
experienced great
contrasts during his
lifetime. He was born
into a family with a long
lineage of owning and
operating ironworks,
and he himself acquired
Engelsberg at a young
age through inheritance
and purchases.

As the son of a dedicated
natural history collector,
it was not unexpected
that Paul August Timm
(1831–78) continued in
the tradition. Like his
father, Gabriel Casper, he
also studied in Uppsala
and succeeded him as
the owner and proprietor
of Engelsberg Ironworks.

Gabriel Casper married
his cousin Mathilda
Lovisa, née Nettleblad
(1800–92) in 1827.
Whether Mathilda
Lovisa was involved
in the collection
of natural history
objects is not known.

when it was exhibited in Stockholm, years later. Even though their natural history collection was unusually extensive and in many ways spectacular, one must at the same time keep in mind that it was just one of many such examples. If Gabriel Casper had been influenced by the Grill family, it was most likely as part of a larger network of collectors and enthusiasts.

Another example of a collector who can be compared to the Timms is Johan Lorentz Aschan (1772–1856), the ironworks owner and doctor at Ohs järnbruk (Ohs Ironworks) in Småland, who left behind a long overlooked collection of minerals and was also connected with their friend Jöns Jacob Berzelius (1779–1848). Aschan himself had defended his dissertation in medicine and botany at Uppsala University and was thereby a fully-fledged scientist with officially recognised merits. His interest in collecting minerals was not just a hobby; he owned mines and made attempts to extract copper from copper ore (chalcopyrite). His collection even contained minerals from exotic places around the world, which may not have had a direct connection with his own operation. Both useful and curious things coexisted in the Aschan collection, as in many others.

As an era, the nineteenth century is overshadowed by the eighteenth for obvious reasons, both in terms of research and the state of general knowledge. None of the scientists of this time became as famous and revered as Linnaeus, even though there was no shortage of names that definitely put different elements of Swedish natural sciences on the world map.

Berzelius was one such name, and several of Linnaeus's disciples continued to attract attention into the early nineteenth century. Even somewhat later names, such as botanist and mycologist Elias Fries (1794–1878) and the naturalist Carl Adolph Agardh (1785–1859), a professor of botany and natural sciences who specialised in algae, were famous scientists in their day, but are now little known. What remains clear is that the collection of objects of natural history was a very important and in some ways fundamental element in the thinking of that time, having both differences and similarities with the eighteenth century.

In the book *Vetenskap i provinsen* ('Science in the Provinces', 1999), the historian of ideas Jakob Christensson discusses the reasons why so many people were drawn to the laborious task of collecting, instead of enjoying other easier forms of leisure. He suggests that

the excitement of searching and the joy of discovery, the teamwork and friendships that developed throughout the process and through the correspondence with like-minded people, the collecting and cataloging was as rewarding as it was time consuming, the writing, reflection and eureka moments of how everything was connected…were some of the things that made the lives of the amateurs meaningful.

Which was surely true of the Timms as well.

Father and son Timm at Engelsberg Ironworks

After his death in 1870, Gabriel Casper Timm was described as 'a friend of flora not without significant botanical insights'. It was in many ways an understatement. Ever since he was a young man, he had held a burning interest in nature, which, as time went by, resulted in the devoted collecting of plants, insects and shells, stuffed birds, minerals and other things from nature's treasure trove. In his notes that survived him, we learn that, even in his old age, he continued to venture out on botanical excursions and '*hittade rara växter*' ('found rare plants') for his herbarium. And his lifelong passion for collecting was passed on to his son, Paul August. Insects and birds' eggs in particular were supposedly Paul August's speciality, and we know that he went to Norway, among other places, to collect them. A large part of the collection was most likely created through cooperation between father and son, and the wide network of like-minded collectors that they were in contact with. No exact chronology of when the different parts of the collection were compiled, or by whom, has so far been established. However, we do know quite a bit about the general picture and the context of the collection thanks to existing documentation about the family and life at the ironworks.

Gabriel Casper was born into a family with a long lineage of owning and operating ironworks. The family originally came from Germany, and had an ancestor in the 1600s, the Stockholm hat maker Paul Timm, who became interested in mining and iron production and eventually became an ironworks owner. For many generations, members of the Timm family owned and operated the *Sembla bruk* (Sembla iron mill) outside Fagersta in Västmanland, not far from the Engelsberg ironworks. It was where Gabriel Casper was born in 1799, and where he received an introduction to the trade from his father, Anders Timm (1766–1819), the estate owner and proprietor of the ironworks.

Exactly how early on he took an interest in natural history is not known, but he received a solid education and upbringing. According to his student records, Gabriel Casper attended Uppsala University between 1814 and 1820. He received education in a number of subjects, including dance and fencing, but, as far as one can tell, never completed a higher academic degree. He studied history under Erik Gustaf Geijer and read botany with Linnaeus's disciple, Carl Peter Thunberg (1743–1828), which is of particular interest. Is this where his passion for collecting was established? At that time, Thunberg was one of the most famous scientists in his field and has even been called 'the Japanese Linnaeus' and 'the father of South African botany' for his studies of the flora and fauna in these countries. Listening to stories of his adventures and his memories of the great Linnaeus must have left an impression. In Thunberg, Gabriel Casper met someone who had personally known Sweden's most famous natural scientist of all time.

CAROLO PETRO THUNBERG.

Carl Peter Thunberg (1743–1828), a disciple of Linnaeus, was unusually well travelled and internationally known for his pioneering research on Japan and South Africa, among other things. When Gabriel Casper studied under him in Uppsala, they exchanged 'rare flowers'.

H.Osti Upsala

Botanist and mycologist Elias Fries (1794–1878) gained recognition both as a scientist and a cultural personality and belonged to the group of natural scientists who were shaped by the ideals of Romanticism. He is one of the few natural scientists elected to the Swedish Academy.

Following pages: The artist Olof Arborelius (1842–1915) had family connections to the Timm family and founded a vibrant colony of artists who were inspired by the Swedish landscape, such as this example near Engelsberg Ironworks. In 1893 he painted 'Lake view at Engelsberg, Västmanland, a view of Lake Snyten', that would later be renamed *Sverigetavlan.*

During Gabriel Casper's time at university, the ideals of Romanticism flourished, whereby broad general knowledge was highly valued, as were insights into both spiritual and earthly matters. His meeting with the leading minds of Uppsala meant that he had direct contact with these lines of thought, and they seemed to shape many of his future areas of interest. After his studies, he served for a period at *Krigsexpeditionen* (the War Office) in Stockholm. However, running an ironworks was of course a more or less pre-destined career path, which, in 1825, became a reality for Gabriel Casper when he bought his way into and became part-owner of the Engelsberg ironworks.

He had previously inherited shares in the iron mill and, through additional succession settlements and buy-outs, he became sole owner of the entire iron-works only a few years later with his wife, Lovisa Mathilda Nettelbladt (1800–92), whom he had married in 1827. Extensive renovations and investment were made at the ironworks, and the old manor house received a thorough refurbishment, enabling an aristocratic lifestyle and an active social life. Guests arrived in large numbers and were entertained with dining, hunting, walks, skating, parlour games and educated conversations. Their host would also attend agricultural meetings and gatherings of the *Bergshandteringens vänner* (the Society of Metal and Mining Industries), other mining industry meetings, parish meetings, the *Bergsting* (Mining Court of Law), *Jernkontoret* (the Swedish iron and steel producers association) and such associations befitting an ironworks owner. He will surely have met many like-minded people with similar interests.

Collecting was a task that had to be fitted into a very hectic life. While the iron production rumbled on in foundry and forge, and when guests were not visiting, spare time could be dedicated to natural science and reading. One can imagine how father and son – and perhaps other members of the family – would eagerly await reports about new findings and discoveries, and rejoice over rare specimens, either of their own finding or those sent to them by other collectors. Uncommon items would surely have been proudly displayed to a selection of visitors amazed by the flowers and insects from far and wide, knowledgeably presented by Gabriel Casper and, later, Paul August. The pleasure of welcoming like-minded collectors would not have been unusual for them; they were apparently in contact with a great number of them.

Several times a year, Gabriel Casper and Lovisa Mathilda made trips to Stockholm, and sometimes even to Uppsala and other cities. In the capital, they usually stayed with Lovisa Mathilda's sister, Tullia Henriette, who was married to the prosperous snuff factory owner Jacob Fredrik Ljunglöf. He, in turn, was acquainted with Jöns Jacob Berzelius, another famous name with close ties to the Timm family. Apart from of business, these visits to Stockholm were also spent on different kinds of leisure. Book purchases were regularly made, as were trips to the *Spektaklet,* the theatre.

O. Arborelius -1893-

Brukspatron Herr O. Finm

Debet

6 st. Appel träd 1 Rd Bko — 8 —
6 st. Päron Dito — 6 —
3 st. Körsbärs träd — 3 —
3 st. Plommon träd 1 Rd 16 s st. — 4 —
6 st. ägta röda Provenser 1 Rd Bko — 6 —
6 st. Röda Engl. Stickelbärsbuskar — 2 —
2 st. Röda Svenska Vinbärsbusk: — 1 —
2 st. Ribes aureum 12 s. st. — " 24
2 st. Lonicera Caprifolium — " 24
3 st. Amygdalus nanna — " 36
2 st. Pyrus baccata — 2 —
2 st. Pyramid poppler — 1 —
2 st. Balsam Dito — 1 —
2 st. Sillver poppler 32 s. st. — 1 16
6 st. Lilium Candidum hvit lilior — 2 —
2 st. Fritillaria Imperiales — " 32
4 st. Lilium martagon — — —

Summa Banco Rd 39 — 36
3 st. mattor och segelgarn m.m. till Emb: 1 — 24

Bergianska Trädgården den 14 Oct: 1829

Betalt som Qviteras
C. M. Sundström   N. Lundström

The trips also provided an opportunity to expand their scientific knowledge. For example, in July 1851, 'solar eclipse glass' was noted on the list of expenses, as was '*Luftballongs visande*' ('the display of hot air balloons'), along with an entrance fee for '*naturforskaremötet, hvarunder jag mest uppehöll mig uti Botaniska sectionen*' ('the nature science meeting during which I mostly kept to the botanical section').

It was also here that he met with Elias Fries, who donated several plants to the herbarium at Engelsberg and whose writings were of great significance for Gabriel Casper. In another year, 1841, they occupied themselves with '*beseende af orangeri, Naturkabinett, kyrka, Konstkabinett m.m*' ('the viewing of an orangery, nature cabinet, church, art cabinet, etc'), in Uppsala. Purchases of flower bulbs and seedlings were often made and taken back to their garden at the ironworks, where they also had an orangery.

A visit to the Timms at Engelsberg was well worth the journey for those who had an interest in natural history or nature in general. Located in the woodlands between Lakes Snyten and Åmänningen, Engelsberg had long been recognised for its natural beauty and species-rich environment. The travel writer Otto Sebastian von Unge (1797–1849), who was also from Västmanland, described the area poetically in 1829: 'The walking path between Engelsberg and Högfors, across rose studded fields, over streams and leaf covered heights...is refreshing for the wanderer who comes from the plains, where every patch of land appears to be factory produced.' von Unge continues by asserting that the Nordic summer here is 'as beautiful as a wedding day', its unspoiled landscape standing in stark contrast to the barrenness of the worked farmlands, a comparison typical of the mindset of Romanticism.

Towards the end of the nineteenth century, the area would attract a fairly large number of artists who took inspiration from the natural world. An artists' colony emerged, led by Olof Arborelius (1842–1915), who was related to the Timm family. Arborelius spent no less than 35 summers at the Engelsberg ironworks, renting one of the wings of the stately home for a period of time. Another member of the group was Arvid Mauritz Lindström (1849–1923), who, in a letter to his friend Carl Larsson, reported that 'this place is beautiful as hell'. The natural world around Engelsberg has left a significant impression in the county's art history, and it remains preserved in its physical form in the herbarium.

Paul August Timm inherited both his interest in nature and his profession from his father. He too received a very thorough education, firstly through homeschooling with a private tutor and later through further studies in Uppsala. He found the world of insects particularly appealing, and, according to some sources, is accredited with being responsible for the butterflies and parts of the insect collection. He married Gabriella Catharina Arosenius (1835–1920) and had a son, Clas Gabriel. Paul August died at the age of 47 in

Accounts from the Timm family are preserved in the archives at the ironworks. They give an account of everyday life as well as insight into the major renovations that Gabriel Casper commissioned during his first years as owner. Here is a list of tree varieties purchased from the Bergius Botanic Garden, dated 1829. Various kinds of fruit trees, poplars and berry bushes were planted in the manor park, which also had an orangery.

1878, surviving his father by a mere eight years, and it is hard to know what happened to the collection during that relatively short interval.

Preserved letters reveal that Paul August was in contact with other collectors, who he occasionally offered doubles to, and his trip to Norway indicates that he truly put a lot of time and effort into his collecting. Meanwhile, the ironworks suffered very hard times during this period – what has been called *bruksdöden* ('the great mill death'), when many small and medium-sized iron mills could no longer compete with the mass production of larger operators, hit the Engelsberg ironworks with full-force in the 1860s. The outcome was so severe that the iron mill faced bankruptcy in 1869, and two thirds of it was sold at executive auction under very sad circumstances.

The share of the estate that Paul August inherited consisted of the manor house and a minor share of production output – which was apparently enough to keep him afloat during the ensuing years. Yet, despite this financial crisis, the family's natural history collection remained undisturbed, and by this time it had come to fill an entire room in the stately home. In the estate inventory recorded after Paul August's death, we learn that everything remained intact in 'the estate-owner's inner room'. In here, the wooden cabinets were placed in a row, along with chairs and glass cases containing stuffed birds. The room also housed shotguns and weapons used for hunting and trips to the forest. In all, the insect, fauna and egg collections were valued at 200 Swedish crowns, and the stuffed birds at 100 crowns, a considerable amount of money for the time. There may have been other collections that were both larger and more valuable, but this was nonetheless an important store. One can speculate as to whether the assessor understood the true worth of the collection, but the value arrived at does in some way represent an understanding that it was of significant merit.

After Paul August's death, the exact fate of the Timm family collection is not known in detail, other than that it was preserved in good condition. His son, Clas Gabriel, who would be the third and last generation of Timms at the iron mill, was only 12 years old when he lost his father. Paul August may not have had the opportunity to initiate his son into the world of natural science in the same way that he himself had once been instructed by his own father. Whether Lovisa Mathilda, or Clas Gabriel's mother, played an active role in the collections and their preservation is another matter. What we do know is that the cabinet remained locked and, as far as one can tell, the Timm collection has lain dormant ever since. In the 1960s, a large cow shed at Engelsberg Ironworks was converted into archive rooms, which is where the cabinets reside today, protected from fire and the fluctuating indoor climate of the main house. Researchers have visited it on a few occasions, but substantial studies of the material have not yet been undertaken. The potential for future research is great.

Opposite and following pages: 'Wall plaques for the teaching of botany' was a series of educational posters published by botanist Nils Johan Andersson (1821–80). The different parts of the plant are presented in great detail, making the study of the real specimens easier.

Page 74–75: 'Prospect of the Grill family's bird collection at Söderfors, and the existing wall opposite the entrance' is how this watercolour is captioned. No less than 162 different birds are visible, neatly arranged by the Grill family aide Pehr Lindroth. The artist was his brother David Johan, who signed it in 1804.

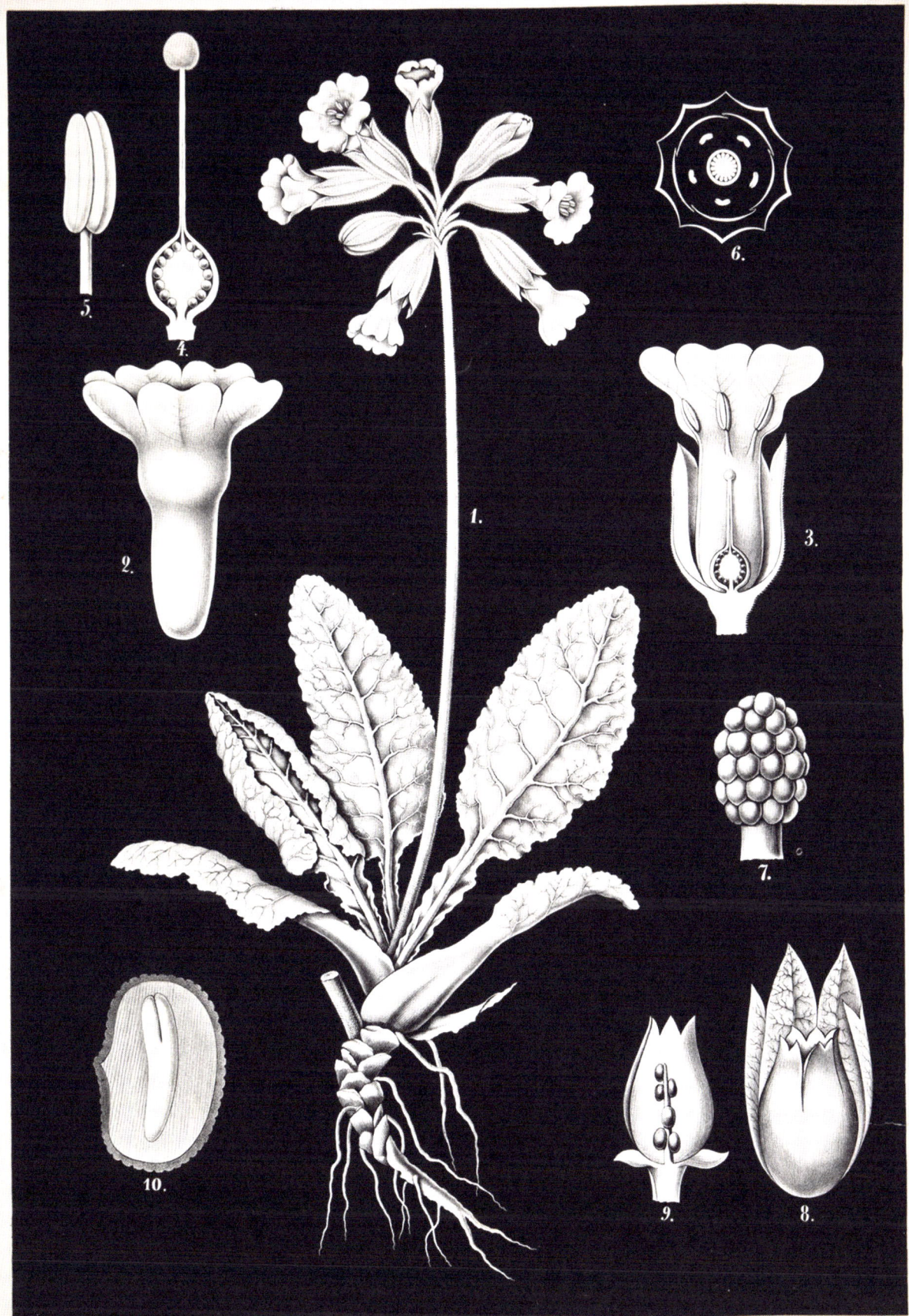

PRIMULA VERIS L. GULLVIFVA.

OXALIS ACETOSELLA L.— HARSYRA.

ORCHIS MACULATA L- YXNE.

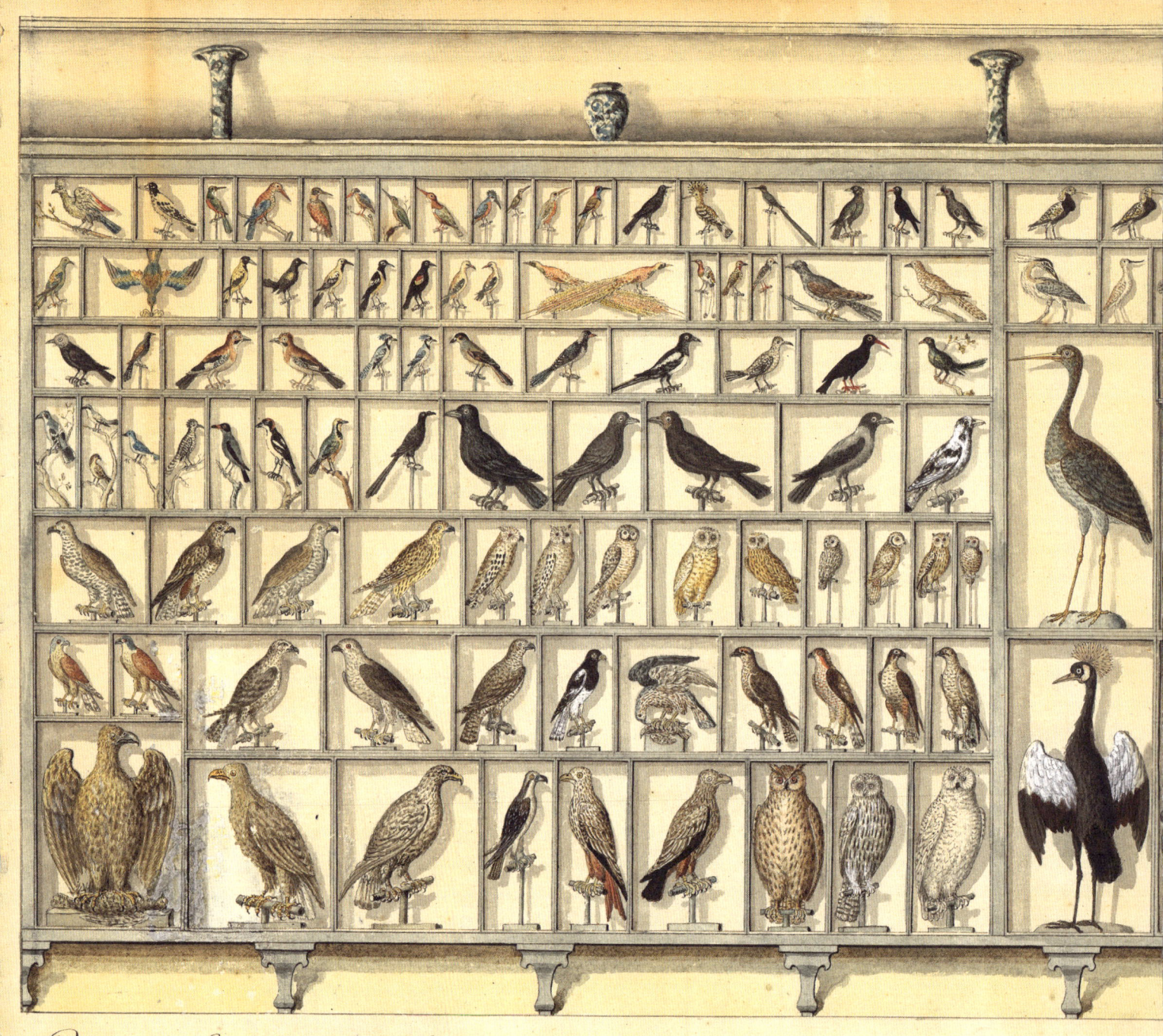

Prospect af Grilliska Fogel=Samlingen på Söderfors, och den mitt emot ingången varand
ferente Foglar, uti dylik ställning uppstoppade och rangerade af Pehr Gustaf

en af 14 alnars längd och 8 alnars högd, i denna ordning beprydd och fullsatt med 162 dif=
roth. Tecknade och ritade af David Johan Lindroth 1804.

The Timms' natural history collection may not have been as artistically or strikingly arranged as the flamboyant *Wunderkammers* of the Renaissance and Baroque periods. At Engelsberg Ironworks, they were satisfied with simple wooden cabinets in the manor's common room. Yet there seemed to be a mutual starting point in the fascination with and devotion to collecting, which was certainly as deeply felt by the Västmanland ironworks owner as it was by European collectors. This illustration is from the pharmacist and collector Ferrante Imperato's Dell'Historia Naturale, printed in Naples in 1599.

RITRATTO DEL MVSEO DI
FERRANTE IMPERATO

There is a risk that the importance of the Timm family's natural history collection could either be exaggerated or understated. Browsing through the leaves of the herbarium and admiring the plants, insects, minerals, bird's eggs and shells makes for a powerful experience; it is truly a monument of a bygone era and a testament to their great devotion and passion for natural history. It raises questions about the Timms and other collectors of that time, and also larger questions about why we humans are gatherers in the first place.

How does the collection rate on a scale measuring for values of recreational pleasure and existential complexity? How did the collection change over time? What did the Timms think about the new scientific findings they heard of, and how did they see their role – as hobbyist amateurs or active participants in the scientific project?

Neither Gabriel Casper nor Paul August Timm left any known written explanation for their collecting, and we therefore know very little about their actual ambitions and intentions. The existence of the collection itself has also not been widely known in this field of research. That it will now be published in book form is therefore rather significant. We cannot know what the Timms themselves thought about the subject – there is no indication that they sought publication, nor did they wish to elevate themselves as scientific authorities. Perhaps they were content with their role as enthusiastic amateurs and occasional informants to science. Yet their surviving collection is a remarkable testimony to the desire for discovery and the deep admiration that many people of the nineteenth century felt for nature and the mysteries of the natural world. Darwin and his theory of evolution had not yet made their breakthrough. The world was not yet demystified or deconstructed, and other ways of thinking would soon set the old intellectual order spinning. Regardless of the interpretations we make today, the collection is a valuable conversation partner, through which its lack of clear answers calls for further in-depth study.

The Timm family collection can therefore be seen as an encapsulation of that moment in history when religious or magical faith and belief were not seen as incompatible with science and the material world – at least not for two knowledge-thirsty ironworks proprietors from Västmanland.

At the top of the Augsburg art cabinet is a Seychelles nut, which at the time was a true rarity. It has been transformed into a boat-shaped trophy raised by the sea god Neptune, as it was believed at the time that they came from a tree at the bottom of the sea. In the boat is Venus, the goddess of love.

Högädle Herr Bruks Patron !

Köping o. d. Björskog
d. 28 Jan. 1840

Jag får tacka på det högsta för sist, för den
ärade bekantskapen, och det nöje som skänktes
mig vid mitt besök på Engelsberg! För brefvet,
som jemte mosshäftet och de båda växifrön, efter
jag äfven mycket tacka. Sedan jag hade nyss
besökt Engelsberg, gjorde jag ännu åtskilliga utvandringar
omkring Norberg, och gjorde äfven ett besök vid
Norn, en bekantskap gjordes med Pastor Strandberg
Åtskilliga ganska goda fynd gjorde jag i Norbergs-
trakten.

Jag har härmed äran afforsända en liten
mossremiss, som innehåller ena mossor jag haft så

Ever since the middle of the nineteenth century, the Timm family's natural history collection has been preserved at Engelsberg Ironworks in Västmanland. The following series of images shows some of the contents of this extensive collection of flowers, herbs, grasses and other plants that have been painstakingly mounted on thin sheets of paper. Both wild and cultivated species can be found here: the magnificent, the inconspicuous; some are common while others are rare. Some originate from places scattered around the Nordic region, while others were collected nearby, in the Västmanland countryside surrounding Engelsberg Ironworks. A number of them come from the ironworks' garden itself that is still maintained onsite to this day. For most of the plants, the species name, collection site, date of collection and collector, have been noted by an elegant hand or with a printed label. It is almost uncanny how well preserved the contents of the herbarium are, as is the rest of the natural history collection. For it would seem that botany was not the only subject that interested Gabriel Casper and Paul August Timm. In other cabinets we find drawers containing brilliantly coloured butterflies, beetles and other insects, shimmering stones and minerals, birds' eggs and seashells. Father and son Timm's collecting even included coins, scientific instruments, stuffed birds, books and other things that served the ideals of education at the time. But the natural objects seem to lie at the core of the collection and became the hub around which the family built a private realm beyond the chores and other a routine aspects of everyday life at the ironworks.

Not many letters or other personal materials
from the Timm family have been recovered
to tell us more about the family's own
thoughts regarding their collecting. However,
the fact that the original cabinets remain is
unusual, and gives us a sense of the original
context in which the collection was created
and assembled.

Växter från Torneå Lapp-
mark, samlade af Hr Rektor
L. Forelius 1864.

Calamagrostis phragmitoides.
    d⁰   lapponica.
Allosurus crispus
Cystopteris montana.
Aira flexuosa β montana.
Poa alpina.
Triticum violaceum.
Agrostis borealis.
Poa cæsia.
Aira alpina.
Hierochloa alpina.
Festuca ovina vivipara.
Avena subspicata.
Phleum alpinum.
Carex tenella.
    d⁰ Persoonii.
    d⁰ rupestris.
    d⁰ chordorhiza.
    d⁰ canescens β subloliacea.
    d⁰ dioica.
    d⁰ tenuiflora.
    d⁰ capitata.

Carex festiva.
d.⁻ rigida.
d.⁻ lagopina.
d.⁻ irrigua.
d.⁻ saxatilis.
d.⁻ ampullacea β borealis.
d.⁻ loliacea.
d.⁻ Buxbaumii.
d.⁻ atrata.
d.⁻ Vahlii.
d.⁻ ustulata.
d.⁻ pedata.

Eriophorum russeolum
d.⁻ Scheuchzeri
Sparganium hyperboreum.
Luzula arcuata.
d.⁻ spicata.
d.⁻ parviflora.
d.⁻ Wahlenbergii.
Juncus triglumis.
d.⁻ biglumis.
d.⁻ alpinus γ uniceps
d.⁻ trifidus.
d.⁻ balticus
Nardus variegatum scirpoides
d.⁻ d.⁻

Different types of records, lists, notes and more are preserved in the collection. Everything testifies to an ongoing process of making new discoveries and having new reflections.

Following pages:
Some people printed and bound catalogues of their natural history collections. This was never the case for the Timm family collection, even though the depth of their interest cannot go unnoticed. On the left: a bundle of notes carefully tied together with a delicate string, perhaps in anticipation of a rewrite that was never made. On the right: a transcript of a lecture by Elias Fries from 1839, written in elegant penmanship by Gabriel Casper.

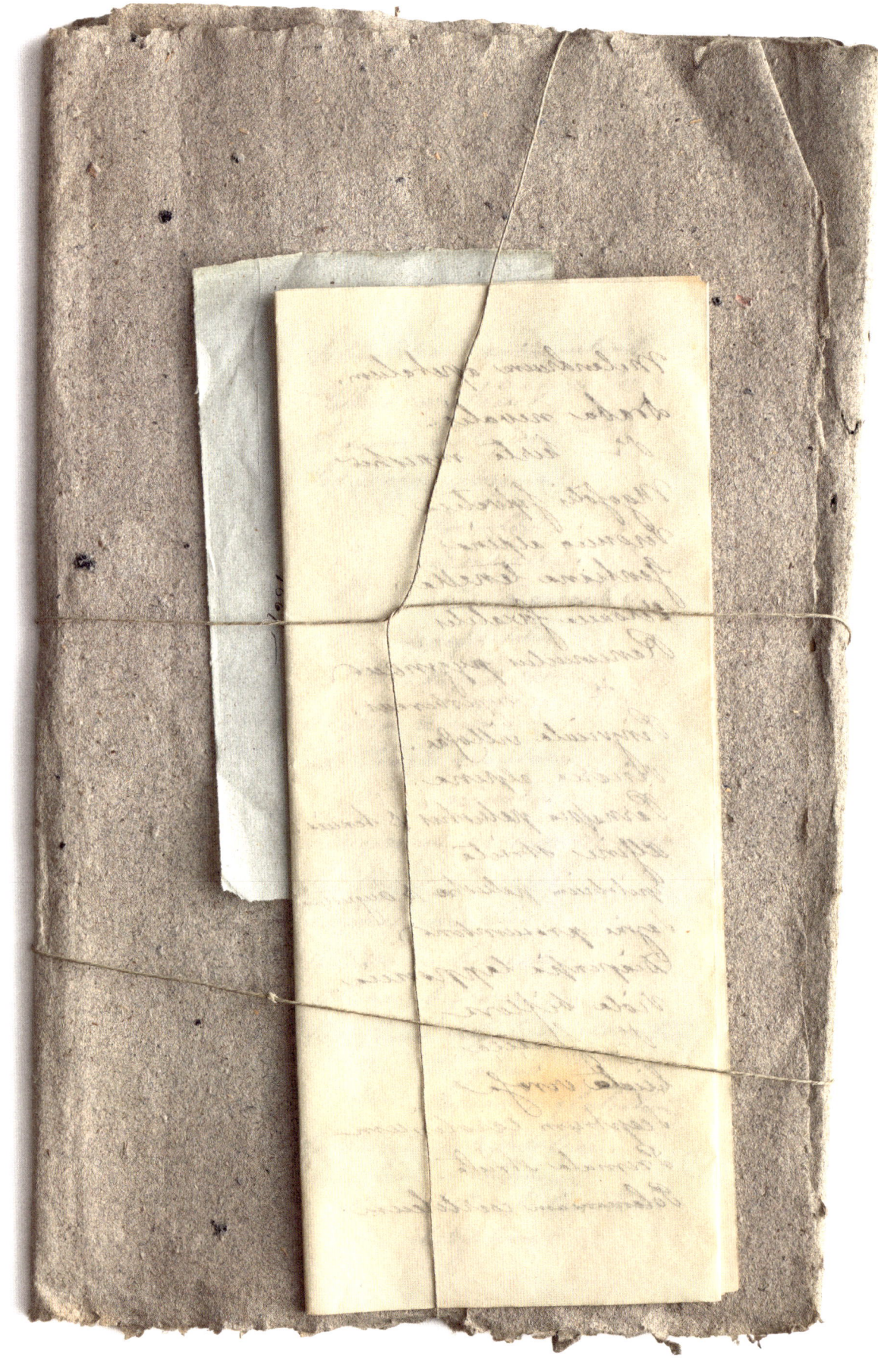

Wäxt-Biologie.

...Biologie är Läran om Wäxternes Lif...

...frågan derwid är: "Hvad är Lif?"...

...ej definieras, utan endast till sina...

...beskrifvas. Man har uppgifvit...

...willig, fö. hos Djuren, forsilig fö. hos...

...lefvande väsende, men rätta en Natur...

...kalla de Naturens...

...uppstå förändringar, som ofta for...

...Magnetism... Affiniteter...

A number of cultivated plants from the garden remain in the herbarium. The fact that accounts of the construction and management of the garden have also been preserved makes the Timm family collection unique. Here are lists of some of the seeds that were purchased.

Frö Lista

	Banco	
6. lod. lägsta miig Hvitkål	"	24
4. lod Sockerrot kåls frö	"	18
½ lod Tidig Blomkål	"	16
½ lod kål Rabij	"	3
4 tt Åker rofvor	"	24
2. lod Gottlands rofvor	"	6
1. lod Turnipps rofvor	"	3
4 lod morots frö	"	12
2. lod Röbeta	"	6
2. lod Radis	"	6
1. lod Rättich	"	3
3. lod Rot persilja	"	9
1. lod Krusperssilje	"	3
4 tt Sabel Ärter	2	"
3 tt Caprosiner Ärter	1	24
6 tt Walska bönor	1	24
2 tt Rosen bönor	"	16
2 tt Swarta stångbönor	"	16
2 tt Gula krypbönor	"	8
2 tt vita krypbönor	"	12
1 lod purjo & 1 lod Hufvud Sallat	"	16
6 lod Rödlök ½ lod Tysk Salvia	1	2
½ lod vit Spansk lök		16
½ lod mejram	"	6
½ lod Basilika	"	2
2 tt Rund Spenate frö		32
några Melon och Candeloupe kiärnor		12
Diverse Blomster frön	Transp: 11	31

Gabrielsson

Mysk luktende Geranium —
Engelberge Tratn 1841.

Bidens grandiflora
Engelbo. Tratn. 1841.

Lysimachia nummularia
L.
Stockholm. Rålambshof
E W Johanson

Pulsatilla vulgaris L.

Upsala

C. U. Johanson

Verbena aubletia

Guldberg Vrad? 1841

Campanula medium ?

Engelb. Trädg. 1841.

Senecis saracenicus
Engelsborg 18...

Cypripedium Calceolus L.

Rosenlani

Fritillaria Meleagris
Upsala
Thunberg

Herbier J.-E. Zetterstedt.
Plantes Pyrénéennes.
Aspidium Oreopteris Sw.
Zett. Pl. Pyr. sp. 1447.
Pyrénées centrales, Superbagnères.
$^{25}/_8$ et $^{20}/_9$ 1856.
J.-E. Zetterstedt.

Tulipa silvestris
Gottf. Sichling.
P. C. Apelius

Angelica litoralis.
Sthlm. Skum Sund.

C W Johanson

110

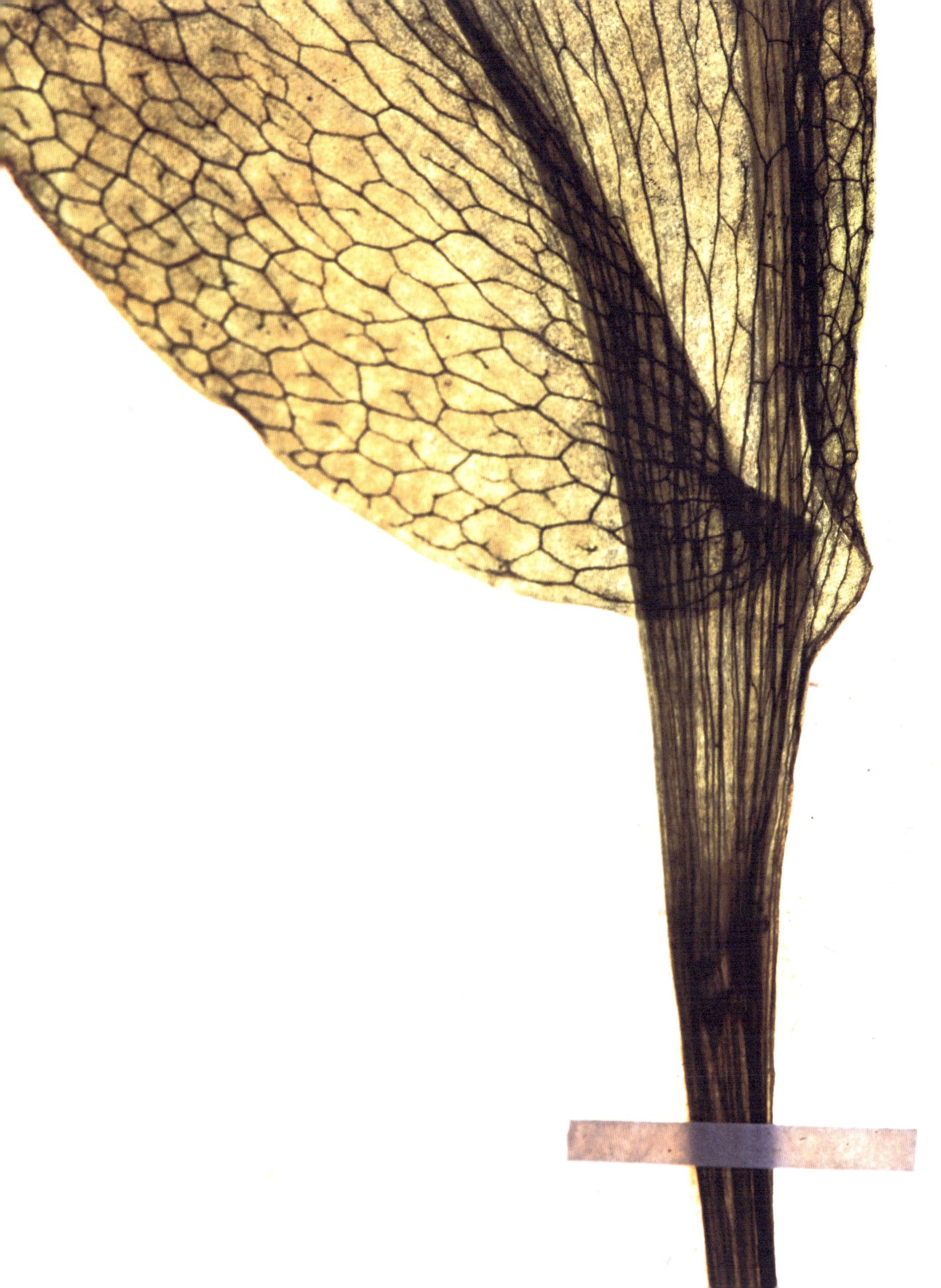

Blechnum spicant.
Roth
Gotl. Wisby
P. C. Afzelius

Blechnum spicant Roth
Småland Femsjö 1851
E. Fries

Cypripedium Calceolus L.

Rosenlani

120

Lotus corniculatus L.
Westre Engelborg
Petiium

124

Asplenum . 2.

Aspidium montanum dw.

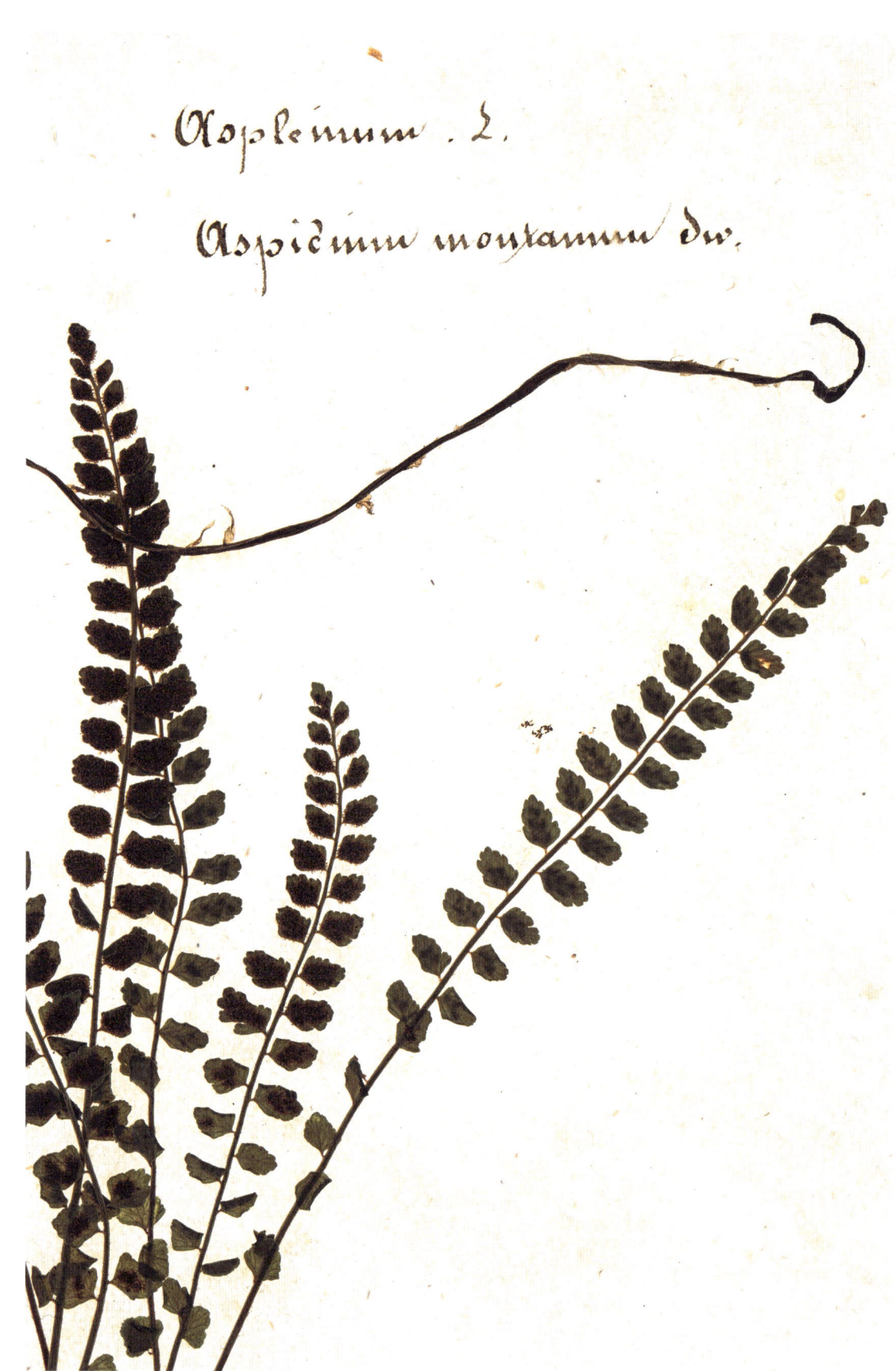

Butomus umbellatus L.
Skåne Åhus
Gadamer

129

137

Cabinet containing eggs
and other natural objects
at Engelsberg Ironworks.

Parus ater Linn.
Tota?,
A.J.F. Wallengren

Parus cristat...
Dalen
Smaland ...
A.J.F. Wa...

Mauda cristata
Tyskland
mb mus...

cristata

...gårdsmyg...
Engelb. 5 Juni
P...

Fring...
...
N...

...gen
modhagen
Nordlanda

Pilgrims Falk.

Falco pereg...
Djupnäs Skog 5.8 Maj 1870
P. A. Sieverin

Falk.

Dvärg Falk.

Fiskjuse.

The insect cabinet at
Engelsberg Ironworks.

Polyphylla

fllo

olontha Fabr.

lgaris Gyll.

Oryctes Illiger.

Anisotoma Illig.	badia Stm.	castanea Er.
dubia Illig.		
	parvula Schltr.	orbicularis Er.
Triepkei Schm.		
arctica Thoms.	Cyrtusa Er.	Amphicyllis Er.
	subtestacea Er.	globus Er.
calcarata Er.		
	minuta Er.	globiformis
ovalis Er.		
	Leiodes Latr.	Typhoceble Thoms.
punctulata Gyll.	humeralis Er.	atra Thoms.
puncticollis Thoms.	axillaris Er.	seminulum Lin.
ciliaris Schm.	glabra Er.	badia Er.
parva Er.	serricornis Er.	laevigata Er.

Agathidium Illig.
nigripenne Illig.
marginatum Strm.
plagiatum Er.
mandibulare Strm.
varians Er.
rotundatum Er.
piceum Er.
nigrinum Strm.
arcticum Thoms.

Eucinetus Germ.
hæmorrhoidalis Germ.
Choleva Latr.
angustata Fabr.
agilis Illig.
Catops Payk.
picipes Er.
tristis Er.
rotundicollis Kellr.
pilicornis F.
morio Fabr.

nigrita Er.
nigricans Gyll.
flavicornis Thoms.
fuscus Er.
femoralis Thoms.
umbrinus Er.
scitulus Er.
Sciodrepa Thoms.
fumata Sp.Lw.
scitula Er.
alpina Gyll.

maculatus *Sycophanta*

brevis

sericea

nitens

reticulata

The mineral cabinet at
Engelsberg Ironworks.

Stuffed birds at Engelsberg
Ironworks.

Image Rights

p. 8: © Engelsbergs Arkiv/Nordstjernan AB, 2021.
   Photo: Lena Granefelt
p. 10: Universitetsbiblioteket, Lunds universitet/
   Public Domain
p. 11: Internet Archive/Getty Research Institute
p. 12: Skoklosters slott. Photo: Samuel Uhrdin
p. 13: The J. Paul Getty Museum, Los Angeles
p. 14–15: Uppsala universitetsbibliotek/Public Domain
p. 16: The Walters Art Museum, Baltimore/CC0
p. 17: TT Nyhetsbyrån/Album
p. 18: Internet Archive/John Carter Brown Library
p. 19: Permission of the Linnean Society of London
p. 20: TT Nyhetsbyrån/Kevin Webb/Science Photo Library
p. 21 (left): Nationalmuseum
p. 21 (right): Uppsala universitetsbibliotek/Public Domain
p. 22: Permission of the Linnean Society of London
p. 23: Bridgeman Images/Look and Learn
p. 26: National Library Australia
p. 28: Biodiversity Heritage Library/Public Domain
p. 29: © Bridgeman Images/Natural History Museum, London
p. 30: © RMN-Grand Palais (Musée du Louvre)/
   Jean-Gilles Berizzi
p. 32: © Reading Museum (Reading Borough Council).
   All rights reserved
p. 33: Alamy Stock Photo/Amoret Tanner
p. 34: Wellcome Collection/Public Domain
p. 36: © Engelsbergs Arkiv/Nordstjernan AB, 2021.
   Photo: Lena Granefelt
p. 40: Internet Archive/University of California Libraries
p. 42: DeAgostini/Getty Images
p. 43: © Bridgeman Images
p. 44–45: Alamy Stock Photo/Artefact
p. 46–47: Rijksmuseum, Amsterdam
p. 48: The Metropolitan Museum, New York
p. 49: © Bridgeman Images/Mitchell Library, State Library of
   New South Wales
p. 50–51: © KHM-Museumsverband, Wien
p. 52: Illustrerad Tidning
p. 54: © Engelsbergs Arkiv/Nordstjernan AB, 2021.
   Photo: Lena Granefelt
p. 56: Gustavianum, Uppsala universitetsmuseum
p. 57: Nationalmuseum
p. 58: © Kungl. Hovstaterna. Photo: Alexis Daflos
p. 59: Naturhistoriska riksmuseet
p. 60: Uppsala universitetsbibliotek/Public Domain
p. 62: © Engelsbergs Arkiv/Nordstjernan AB, 2021.
   Photo: Bengt Wanselius
p. 64–65: Uppsala universitetsbibliotek/Public Domain
p. 66–67: Nationalmuseum
p. 68: © Engelsbergs Arkiv/Nordstjernan AB, 2021.
   Photo: Lena Granefelt

p. 71–73: © Engelsbergs Arkiv/Nordstjernan AB, 2021.
   Photo: Mia Törnberg
p. 74–75: Uppsala universitetsbibliotek/Public Domain
p. 76–77: Universitetsbiblioteket, Lunds universitet
p. 79: Gustavianum, Uppsala universitetsmuseum
p. 80: © Engelsbergs Arkiv/Nordstjernan AB, 2021.
   Photo: Andreas Lisstrand/Italgraf Media
p. 82–83: © Engelsbergs Arkiv/Nordstjernan AB, 2021.
   Photo: Lena Granefelt
p. 84–86: © Engelsbergs Arkiv/Nordstjernan AB, 2021.
   Photo: Andreas Lisstrand/Italgraf Media
p. 87–89: © Engelsbergs Arkiv/Nordstjernan AB, 2021.
   Photo: Lena Granefelt
p. 91–93: © Engelsbergs Arkiv/Nordstjernan AB, 2021.
   Photo: Andreas Lisstrand/Italgraf Media
p. 94–95: © Engelsbergs Arkiv/Nordstjernan AB, 2021.
   Photo: Lena Granefelt
p. 97–100: © Engelsbergs Arkiv/Nordstjernan AB, 2021.
   Photo: Andreas Lisstrand/Italgraf Media
p. 103–105: © Engelsbergs Arkiv/Nordstjernan AB, 2021.
   Photo: Lena Granefelt
p. 106: © Engelsbergs Arkiv/Nordstjernan AB, 2021.
   Photo: Andreas Lisstrand/Italgraf Media
p. 107–163: © Engelsbergs Arkiv/Nordstjernan AB, 2021.
   Photo: Lena Granefelt

Index

COLLECTING NATURE

A History of the Herbarium and Natural Specimens

Published by Bokförlaget Stolpe, Stockholm, Sweden, 2022

© The authors and Bokförlaget Stolpe 2022,
in association with Axel and Margaret Ax:son Johnson Foundation
for Public Benefit

Clive Aslet is an award-winning architectural historian, author and
journalist, and was for many years the editor of *Country Life* magazine.
Svante Helmbaek Tirén is a writer and curator with a focus on the
history of architecture and design.

Translation of Svante Helmbaek Tirén's text: Fern Scott Olsson
Cover image: Photographer Lena Granefelt
Text editor: Andrew Mackenzie
Design: Patric Leo
Layout: Petra Ahston Inkapööl
Prepress and print coordinator: Italgraf Media AB, Sweden
Print: Narayana Press, Denmark, 2022
First edition, first printing

ISBN: 978-91-89425-64-4

BOKFÖRLAGET STOLPE

AXEL AND MARGARET AX:SON JOHNSON
FOUNDATION FOR PUBLIC BENEFIT